Ion Transporters - From Basic Properties to Medical Treatment

Edited by Zuzana Sevcikova Tomaskova

Published in London, United Kingdom

Ion Transporters - From Basic Properties to Medical Treatment
http://dx.doi.org/10.5772/intechopen.95146
Edited by Zuzana Sevcikova Tomaskova

Contributors
Min Luo, Keerti Mishra, Daniel Balleza, Abigail García-Morales, Aylin López-Palestino, Jun Feng, Katherine Zhong, Shawn Kant, Frank Sellke, Sari Susanna Tojkander, Rosanna Di Paola, Marika Cordaro, Salvatore Cuzzocrea, Fabien P. Pascal Chevalier, Bérengère Fromy, Lisa S. Martin, Emma Fraillon, Anuj Tripathi, Smita Misra

First published in London, United Kingdom, 2022 by IntechOpen
IntechOpen is the global imprint of INTECHOPEN LIMITED, registered in England and Wales, registration number: 11086078, 5 Princes Gate Court, London, SW7 2QJ, United Kingdom

British Library Cataloguing-in-Publication Data
A catalogue record for this book is available from the British Library

Additional hard and PDF copies can be obtained from orders@intechopen.com

Ion Transporters - From Basic Properties to Medical Treatment
Edited by Zuzana Sevcikova Tomaskova
p. cm.

This title is part of the Biochemistry Book Series, Volume 38
Topic: Cell and Molecular Biology
Series Editor: Miroslav Blumenberg
Topic Editor: Rosa María Martínez-Espinosa

Print ISBN 978-1-80355-549-2
Online ISBN 978-1-80355-550-8
eBook (PDF) ISBN 978-1-80355-551-5
ISSN 2632-0983

IntechOpen Book Series

Biochemistry

Volume 38

Aims and Scope of the Series

Biochemistry, the study of chemical transformations occurring within living organisms, impacts all of the life sciences, from molecular crystallography and genetics, to ecology, medicine and population biology. Biochemistry studies macromolecules - proteins, nucleic acids, carbohydrates and lipids –their building blocks, structures, functions and interactions. Much of biochemistry is devoted to enzymes, proteins that catalyze chemical reactions, enzyme structures, mechanisms of action and their roles within cells. Biochemistry also studies small signaling molecules, coenzymes, inhibitors, vitamins and hormones, which play roles in the life process. Biochemical experimentation, besides coopting the methods of classical chemistry, e.g., chromatography, adopted new techniques, e.g., X-ray diffraction, electron microscopy, NMR, radioisotopes, and developed sophisticated microbial genetic tools, e.g., auxotroph mutants and their revertants, fermentation, etc. More recently, biochemistry embraced the 'big data' omics systems.
Initial biochemical studies have been exclusively analytic: dissecting, purifying and examining individual components of a biological system; in exemplary words of Efraim Racker, (1913 –1991) "Don't waste clean thinking on dirty enzymes." Today, however, biochemistry is becoming more agglomerative and comprehensive, setting out to integrate and describe fully a particular biological system. The 'big data' metabolomics can define the complement of small molecules, e.g., in a soil or biofilm sample; proteomics can distinguish all the proteins comprising e.g., serum; metagenomics can identify all the genes in a complex environment e.g., the bovine rumen.
This Biochemistry Series will address both the current research on biomolecules, and the emerging trends with great promise.

Meet the Series Editor

Miroslav Blumenberg, Ph.D., was born in Subotica and received his BSc in Belgrade, Yugoslavia. He completed his Ph.D. at MIT in Organic Chemistry; he followed up his Ph.D. with two postdoctoral study periods at Stanford University. Since 1983, he has been a faculty member of the RO Perelman Department of Dermatology, NYU School of Medicine, where he is codirector of a training grant in cutaneous biology. Dr. Blumenberg's research is focused on the epidermis, expression of keratin genes, transcription profiling, keratinocyte differentiation, inflammatory diseases and cancers, and most recently the effects of the microbiome on the skin. He has published more than 100 peer-reviewed research articles and graduated numerous Ph.D. and postdoctoral students.

Meet the Volume Editor

Zuzana Sevcikova Tomaskova graduated with a degree in Biophysics from Commenius University, Bratislava, Slovakia. In 2009, she obtained her Ph.D. in Biophysics from Pavol Jozef Safarik University, Kosice, Slovakia. She specializes in single-channel current measurement on bilayer lipid membrane and advanced current analysis. She is the recipient of several awards, including a 2010 Young Scientist Award from the Slovak Republic. In 2012, she received a short-term FEBS fellowship and learned the single-channel patch-clamp method at Nencki Institute, Warsaw, Poland. Her work is mainly focused on cardiac mitochondrial chloride channels that play an important role in cardiac arrhythmias. Currently, she is a senior researcher at the Centre of Biosciences, Slovak Academy of Sciences.

Contents

Preface

Ion channels are small proteins that are present in every cellular membrane. They provide the path for ions to follow their concentration gradient and thus transport stimulus or information across the membrane. When talking about ion channels, one cannot forget the counterbalance of the passive currents and maintenance of the ionic gradients mediated by ion pumps (ATPases or ion exchangers).

It is almost 100 years since the first electrical signal in a single squid giant axon was detected by Hodgkin, Huxley, and Katz. During the century, the idea of what is behind the electrical current has moved from a courageous hypothesis to a structure of an ion transporter known on the atomic level and computer simulations indicating the movement of amino acid chains during channel gating or ion pump movement.

Many types of ion channels have been observed, measured, and described since their discovery. There are ion channels so selective that they can distinguish between two small ions like Na+ and K+. In addition, there are ion channels that are permeable for molecules several times larger than ions. These channels differ in the speed of transport, the shape of conductive pores, and gating kinetics. The regulation of ion channel activity involves a broad range of mechanisms. Though the channels vary in many ways, their primary role is the same, which is the controlled passage of ions in a safe way to not disturb cellular homeostasis. Ion channels, regardless of the ion type they conduct, are often the target of medical treatment. Mutations in ion channels that affect channel activity or permeability are behind diseases such as cystic fibrosis, malignant hyperthermia, epilepsy, and many other channelopathies.

This book focuses on the tight connection between an abnormal channel or ion pump function or regulation and the diseases that emerge from these changes. The first section shows the plethora of ion channel types that are involved in pathological processes like cardiac or neurodegenerative diseases. The physiological state is a result of an interplay of many different ion channels. The slightest deviation from the "normal" of any of the small puzzle pieces can evolve into a serious condition. The second section contains chapters on a single ion transporter type and its role in different physiological and pathological states. The third section presents an example of a computer-modeled 3D channel structure that can be extremely helpful in the future of drug development.

Maintenance of the dynamic balance, that relentlessly goes on in every cell, is crucial for human health. I hope that every reader of this book, whether a student or a scientist, will find it entertaining as well as instructive.

Zuzana Sevcikova Tomaskova
Centre of Biosciences,
Slovak Academy of Sciences,
Bratislava, Slovak Republic

Section 1

Different Ion Channels behind One Pathology

Chapter 1

Ion Channels and Neurodegenerative Disease Aging Related

Marika Cordaro, Salvatore Cuzzocrea and Rosanna Di Paola

Abstract

Many neurodegenerative diseases such as Alzheimer's disease, Parkinson's disease, Huntington's disease, multiple sclerosis, amyotrophic lateral sclerosis, and age-related disorders are caused due to altered function or mutation in ion channels. Ion channels are important in maintaining cell homeostasis because they affect membrane potential and play a critical role in neurotransmitter secretion. As a result, it appears that a potential antiaging therapy strategy should consider treating multiple diseases at the same time or focusing on identifying a common target among the biological processes implicated in aging. In this chapter, we will go over some of the fundamental ideas of ion channel function in aging, as well as an overview of how ion channels operate in some of the most common aging-related disorders.

Keywords: aging, ion channels, neurodegeneration, therapeutic targets

1. Introduction

Aging is a natural part of life that comprises both physical and mental changes. In distinct organs, aging occurs at molecular, cellular, and histological levels, including in the central nervous system (CNS) and specifically in the brain [1, 2]. The molecular, chemical, and physical properties of neurons change as we become older, resulting in memory loss, altered behaviors, loss of cognition functions, dementia, and reduced immunological responses. In addition, aging is a major risk factor for neurological diseases including Alzheimer's disease (AD), Parkinson's disease (PD), amyotrophic lateral sclerosis (ALS), multiple sclerosis (MS), and others. Although the basic reasons of aging are unknown, there is widespread agreement that its etiology is multifaceted [3]. Aging theories can be classified into two categories: those that explain aging as the outcome of damage accumulation and those that explain aging as the result of controlled death processes [4]. It is likely that the interaction of these two basic systems influences the aging process, albeit there is a lot of variation across people. Two of the most accredited molecular alteration involved in brain aging are inflammation and oxidative stress that, when happen lead to cells failure. Different studies reported that reactive oxidative species (ROS), and subsequent oxidation of proteins, involved also ion channels [5].

Ion channels are integral membrane proteins that allow the passive diffusion of ions across membranes [2]. In neurons and other excitable cells, the harmonious coordination between the numerous types of ion channels shapes and propagates electrical signals [6]. Understanding the biology of aging mechanisms is essential to the pursuit of brain health. The ability to stratify senior populations and forecast clinical trajectories in pre-symptomatic adult groups could be critical to the future of aging research [4]. In this chapter, we will discuss about the role of ion channels in the brain during aging with particular attention on neurodegenerative disease age-related. Additionally, we will consider if ion channels could be used as future therapeutic targets to decelerate brain aging and age-related pathologies.

2. Brain aging: from physiological to pathological

Scientists have been debating the meaning of aging for a long time. Many people regard aging as an illness in and of itself, while others see it as the gradual loss of function that increases the risk of developing age-related diseases. Scientists view aging as an adaptation to lifelong events, and interventions should support the physiological balance during age-related adaptation, response to acute stress, to avoid disease onset. Adapted capacity in most organs has been shown to occur from the third and fourth decades of life [4]. Aging is a complicated and multifaceted condition marked by a steady decline in physiological and behavioral abilities. Aging happens in all organs at all levels, in the brain [2]. The molecular, chemical, and physical properties of neurons change as we become older, resulting in memory loss, altered behaviors, loss of cognition functions, dementia, and reduced immunological responses. Rather than significant rates of neuron loss, brain aging has been linked to subtle changes in the structure and function of neurons in specific neural circuits. The aging brain compensates for the loss of neurons by growing dendritic arbors and synaptic connections. Dendritic arbors and synaptic connections are lost in the brain in age-related neurodegenerative disorders. As a result, it is unable to compensate for the loss of neurons [7]. Synaptic degeneration, dendritic regression in pyramidal neurons, deposition of fluorescent pigments, cytoskeletal abnormalities, a reduction of striatal dopamine receptors, and astrogliosis and microgliosis are all prevalent features of brain aging in mammals [8]. Despite the discovery of brain aging characteristics in multiple neural networks, the chemical pathways responsible remain unknown [9]. Oxidative stress, inflammation, and ion channel failure are the most widely accepted theories for the development of age-related neurodegenerative diseases [10].

2.1 Oxidative stress in brain

In the 1950s, Harman's free radical theory of aging suggested that reactive oxygen and nitrogen species (ROS and RNS) cause oxidative damage in cellular macromolecules, including DNA, proteins, and lipids, leading to decreased biochemical and physiological function through aging [11]. The changes in phospholipid composition show that ROS-induced lipid peroxidation occurs in the brains of elderly humans and animals with CNS dysfunction, such as cognitive impairment. Furthermore, increased formation of malondialdehyde (MDA) in the brain has been postulated as a symptom of aging [12]. Superoxide anions produced by the respiratory chain and various oxidases, hydroxyl radical created by the hydrogen peroxide interaction with Cu^{+} or Fe^{2+}, and NO produced in response to elevated intracellular Ca^{2+} levels are just two of the most common examples of ROS in neurons [13]. During brain aging, enhanced ROS generation and decreased antioxidants result in redox imbalance, causing age-related disorders. NO-dependent oxidative damage promotes apoptosis in motor neurons. It causes vascular cognitive impairment through the aging of the cerebral cortex [14]. The action of several enzymatic and non-enzymatic systems with cellular detoxification functions, collectively referred to as antioxidants, mediates the hemostasis of intracellular ROS and RNS. The nuclear factor erythroid 2-related factor 2 (Nrf- 2) is the main transcription factor and one of the primary regulators of the antioxidant signaling, such as transcription of endogenous antioxidant enzymes including glutathione (GSH), glutathione reductase (GR), glutathione peroxidase (GPx), catalase (CAT), superoxide dismutase (SOD), and heme oxygenase-1 (HO-1). This antioxidant system is collectively the primary defense system that neutralizes ROS generation inside the cells [15]. Additionally, another major antioxidant defense complexes are the heat-shock response (HSR), a cellular response that elevates the number of molecular chaperones to diminish the adverse effects on proteins caused via stressors, oxidative stress, increased temperatures, and heavy metals. Increased stress tolerance and cellular protection against neuronal injury can be achieved by activating heat-shock protein (HSP) synthesis [16]. As a result, in metabolic disturbances such as age-related neurodegenerative diseases and aging, the heat-shock response plays a critical role in creating a cytoprotective environment [2].

2.2 Inflammation in brain

Another key pathway directly involved in brain aging is represented by inflammation. The immune system is one of the most pivotal protective physiological systems of the organism [17]. Immunosenescence is a concept that describes how aging affects the immune system's function [18]. The participation of senescent cells in host immunity is associated with the release of pro-inflammatory cytokines. This phenomenon is defined as senescence-associated secretory phenotype (SASP). Due to SASP's pro-inflammatory tendency, cellular senescence in various organs and tissues significantly increases inflammation in the aged [19]. NF-κB in response to oncogenic stress and DNA damage initiates the transcription of a host of genes including tumor necrosis factor-α (TNF-α), interleukin-6 (IL-6), IL-8, IL-1β over stimulating SASP [20]. NF-κB is a transcription factor that is induced by inflammatory mediators and reactive oxygen species (ROS) and contributes to both detrimental and protective responses, depending on the types of induction that lead to the co-activation of distinct pathways. In addition, NF-kB activates genes that control cell survival, specialization, inflammatory processes, proliferation, and apoptosis [20]. It has been shown that the age-induced increase of pro-inflammatory markers (CRP, IL-6, IL-1β, TNF-α) is associated with cognitive decline [21]. Microglia are the brain resident macrophages providing its innate immune defense. Microglia, a kind of glial cell, arise from erythro-myeloid precursors in the yolk sac, which inter the CNS during development [22, 23]. In the neurological system, microglia play two roles. Microglia are ramified cells with extremely motile processes that continually scan the brain parenchyma in reaction to hazardous substances, neuronal cell damage, or infections in a healthy adult brain. Microglia have a dual function in the aging process. Microglia, on the one hand, release trophic factors and control cytokines [24, 25]. On the other hand, microglia enhanced amounts of an intricate set of mediators, such as TNFα, TGFβ, and IL1β, which are enhanced in elderly individuals [22, 23]. There has been evidence of a link between neuroinflammatory activation of microglia and neuronal loss, as well as impaired neurobehavioral function and cognitive impairment. Redox sensors found in receptors, transcription factors, and enzymes provide complex communication with oxidizing agents during neuroinflammation. These variables have an impact on the link between neurons and glia, as well as neuronal function, which leads to neurodegenerative alterations [26, 27]. Microglial cells also express a stimulable type of NOS following activation and produce large quantities of NO, which causes oxidative damage to neurons. In neurodegenerative illnesses and brain aging, improper immune cell activation causes functional impairment and synaptic degeneration; when properly controlled, these same cascades play critical roles in neuronal stress tolerance and neuroplasticity. For instance, TNF-α plays a pivotal role in learning, memory, and synaptic plasticity in the hippocampus [28]. Also, astrocytes may potentially play a role in adapting to age-related neuronal stress. These cells clear glutamate from synapses, produce neurotrophic factors and boost neuronal bioenergetic activity. Aging may decrease these astrocyte activities, hence, exacerbating pathogenic neuroinflammatory processes [28–30]. TNFα activates NF-κB which protects cells against neurotoxicity β-amyloid (Aβ)-induced and this activation is required for neuronal survival. NF-κB also promotes anti-apoptotic responses and protects neurons from ischemia and excitotoxic brain injury [31–36]. Furthermore, through its response to TNF-mediated inflammatory stimuli, NF-κB activation plays a critical role in the start and persistence of inflammation, resulting in the stimulation of various chemokines and cytokines [37–42].

3. Ion channels in the brain: from function to dysfunction

Ion channels are key components of neurons that are responsible for nerve impulse and synaptic transmission triggering (neurotransmitter's release). These channels are divided into two big classes: (I) voltage-gated (Na^+, K^+, Ca^{2+}, Cl^-) and (II) ligand-gated (nicotinic acetylcholine receptors (nAChRs), γ-amino butyric acid (GABA), N-methyl-D-aspartate receptors (NMDARs), ryanodine receptors (RyRs)) that are involved in impulse transmission across the synapses [43]. However, during the last several decades, research has found a number of genetic faults or aberrations in channel-forming genes that are linked to a variety of neurological illnesses, including memory loss, movement difficulties, and neuromuscular disorders [44].

Ion channel protein establishes a pathway for ions such as Na^+, K^+, Ca^{2+}, and Cl^- to flow across the lipid bilayer's impermeable barrier [45]. They are known to play three main important functions in regulating membrane physiology: first of all, they set up membrane potential in cells, in which ion movement across the membrane creates a potential gradient that determines resting potentials and generates action potentials; secondary they are involved in the transmission of electrical signals; they are also involved in maintaining electrolytic balance across the cell membrane to regulate cell volume, and last but not least they play a crucial role in the generation of regulatory signals in the cell [46, 47]. Thanks to alternative splicing, their enormous structural variety from monomeric to heteromeric levels support their large functional diversity. The amplitude and duration of the action potential are shaped differently by each cell type's assembly of ion channels [48, 49]. At the intracellular level, ion channels are also present on the surface of the mitochondria, endoplasmic reticulum, and nuclear membrane [50, 51]. The correct functionality of ion channels is necessary to keep physiological homeostasis in the brain [52].

As a result, ion channels have been implicated in a number of age-related dysfunctions [53]. Because aging is associated with physiological changes in ion channel function, aberrant changes in ionic gradients seem to be the core of age-related deterioration in physiological functioning. With age, functional changes in ion channels lead to clinical phenotypes called channelopathies [54].

3.1 K^+ channels

K^+ channels are the most ubiquitous and heteogeneous family of ion channels expressed in excitable and non-excitable cells (an extensive review on this topic can be found in [55]). K^+channels can be divided into four classes: inwardly rectifying K^+ channels (Kir), voltage-gated K^+ channels (Kv), two-pore K^+ channels (K2P), and Ca^{2+}-activated K^+ channels (KCa) [56]. K^+ channels serve an important physiological function in the signaling mechanisms that lead to neurotransmitter release in neurons. They modulate the resting membrane potential, the repolarization phase of the action potential, and the firing frequency to govern neuronal excitability. Given the importance of K^+ channels in so many cellular functions, it's no surprise that changes in their activity have been linked to the development of a variety of neurodegenerative diseases [57, 58]. Furthermore, in recent years, it has been demonstrated that the apoptotic process, which is the key mechanism for cell selection and death in the CNS associated with physiological aging as well as a variety of neuropathological disorders, is critically dependent on changes in ion homeostasis within neuronal cells [59]. K^+ efflux, which results in a drop in intracellular K^+ concentration, maybe a key cause of apoptosis. In fact, in various neuronal populations undergoing apoptosis, an increase in outward K^+ currents have been seen as well as it has been demonstrated that apoptosis has been shown to be inhibited by voltage-gated K^+ channel blockers, whereas heterologous production of inwardly-rectifying K^+ channels has been shown to increase apoptosis in cultured hippocampus neurons [60].

3.2 Ca^{2+} channels

Ca^{2+} is the major trigger of neurotransmitter release, a process that has been thoroughly investigated over the past decades [61–63]. Moreover, it has also become clear that Ca^{2+} is essential for a variety of other neuronal functions, including neuronal excitability, integration of electrical signals, synaptic plasticity, gene expression, metabolism, and programmed cell death [64]. Given its central role in processes that are fundamental to the excitable nature of neurons, Ca^{2+} homeostasis is tightly regulated in these cells. Plasma membrane Ca^{2+} channels allow the passive influx of calcium ions down their electrochemical gradient. These channels are divided into two groups based on the mechanism that controls their transition between open and closed conformations: voltage-gated Ca^{2+} channels (VOCC) and ligand-gated Ca^{2+} channels. The potential contribution of altered Ca^{2+} homeostasis at least to some aspects of brain aging and neurodegeneration was first put forward by Khachaturian in the 1980s, with the formulation of the "Ca^{2+} hypothesis of aging" [65–67]. Early findings in the field that corroborated this hypothesis examined the major transport pathways of Ca^{2+} during aging and found that at least in some types of neurons, such as the principal cells in the hippocampal CA1 region, there is an increased Ca^{2+} influx mediated by increased VOCC activity in aged neurons [68]. Similarly, Ca^{2+} extrusion through the ATP-driven plasma membrane Ca^{2+} pump (PMCA) was found to be decreased in aged neurons [69]. Following that, the attention switched to the intracellular mechanisms of Ca^{2+} homeostasis and how they degrade with age. The increased release of Ca^{2+} from the endoplasmatic reticulum (ER) stores via both the inositol 3-phosphate (InsP3) and ryanodine receptors (RyR) has been confirmed in several investigations, leading to the suggestion that release from the RyR receptor might be a valuable biomarker of neuronal aging [70]. The high influx of calcium ions into the postsynaptic spine appears to be the crucial event leading to the induction of long-term potentiation (LTP), which is relevant to the function of Ca^{2+} dysregulation in memory loss. Importantly, LTP is inhibited by intracellular Ca^{2+} chelators, whereas LTP is promoted when the postsynaptic cell is Ca^{2+}-loaded [71]. Therefore, it is well established that a significant elevation of postsynaptic Ca^{2+} concentration is both necessary and sufficient for the induction of hippocampal LTP [72]. Ca^{2+} homeostasis changes may be directly responsible for neuronal death in some circumstances. Increased intracellular Ca^{2+} levels can cause severe abnormalities in neurons, eventually leading to neuronal death and degeneration [73]. This process is often specifically mediated or even initiated by the diminished capacity of mitochondria to buffer Ca^{2+}. Given the basic relevance of Ca^{2+} homeostasis in the biology of all cells, it's not unexpected that a growing number of studies demonstrate that unregulated Ca^{2+} plays a role in normal aging as well as a variety of pathological disorders. Given the nervous system's incredible cellular variety, a general message emerging from this research is that Ca^{2+} signaling and homeostasis in the nervous system should be investigated. The Ca^{2+} homeostasis mechanism is equally variable across neurons, according to the demands of each neuronal subtype [62]. The intrinsic variations in morphology, connectivity, proteome, and Ca^{2+} homeostatic mechanism of neurons, taken together, are extremely likely to contribute to the selective sensitivity of diverse neuronal populations to different causes of senescence collectively and synergistically. The more we learn about how Ca^{2+} homeostatic processes interact with distinct neurons' inherent properties, the closer we will be to devising cell-specific therapeutics [62].

3.3 Na^+ channels

Voltage-gated sodium channels (Nav channels) are fundamental for the origination and transmission of signals in electrically excitable tissues. Na^+ channels are abundant in neurons and glia throughout the central nervous system and peripheral nervous system (PNS) [74]. The genesis of neurological disorders, including as idiopathic epilepsy, ataxia, and pain sensitivity, is heavily influenced by mutations in genes encoding Na^+ channels [75]. This is most likely due to changes in the synthesis and/or trafficking of Nav channels, which modify their surface expression and impact the neuron's electrical excitability even while the channel's conducting properties remain unchanged. Changes in the function of voltage-gated Na^+ channels have been observed during the aging process [76]. These alterations were attributed to an age-related reduction in excitability, which is controlled by voltage-gated Na^+ channels. Furthermore, age-related changes in voltage-gated Na^+ channel activity have been proposed as a possible explanation for the decreased excitability seen in skeletal muscle fibers of old rats [77]. Considering the fundamental role of Nav channels in the modulation of neuronal responses during pathophysiological conditions, and the fact that RNS and ROS may play a role in neurodegenerative events, the study of Nav channel modulation by these free radical species assumes a particular pathophysiological relevance. Recent evidence shows that oxidant-induced alterations in the characteristics of Nav channels may play a role in membrane excitability and conductance modulation. Na^+ currents were also elevated when NOS was inhibited or

NO• was scavenged by hemoglobin and ferrous diethyl thiocarbamate. These findings suggest that RNS may act as autocrine regulators of Na^+ currents in these neurons, inhibiting them. NO•, on the other hand, could potentiate the inactivation resistant Nav channels currents (INaP) seen in hippocampus neurons and posterior pituitary nerve terminals [1, 13, 78]. The current carried by these channels appears to be increased not only by the significant rise in NO• levels evoked by NO• donors but also by the lesser increase triggered by constitutive NOS activation [1, 13]. In hippocampal and pituitary neurons, NO• can cause an increase in Na^+ currents, but it has the reverse effect in peripheral neuronal cells. As a result, it appears that NO• might have either neuroprotective or neurodegenerative effects due to its dual effects on various neuronal sodium channel populations. These effects are probably due to the variety of Nav channel subtypes expressed in the CNS and PNS.

4. Ion channels and neurodegeneration

Ion channel deficiencies and/or mutations relate to many forms of neurological diseases. Na^+, K^+, Ca^{2+}, and Cl^- channel subtypes, for example, have been connected to the pathophysiology of dyskinesia, seizures, epilepsy, and ataxia [79]. Following we briefly discuss the role of ion channel modification in the most common neurodegenerative disorders age-related.

4.1 Ion channels and Alzheimer's disease

AD is a kind of dementia marked by cognitive impairment, memory loss, and neuronal death. A buildup of Aβ peptides, tau hyperphosphorylation, and mutations in the catalytic domain of γ secretase are all elements that contribute to the disease's focused characteristic. The ionic imbalance has been linked to AD development, and in particular an aberrant intracellular concentration of Ca^{2+}, Na^+, K^+, and Cl^- [80]. Hardy and Higgins were the first to show that Aβ peptides disrupt Ca^{2+} homeostasis in neurons and increase intracellular Ca^{2+}, which Mattson and his colleagues later validated [81]. Currently, multiple investigations have established the base for a novel concept: Aβ peptide is dangerous to neurons in part because it forms abnormal ion channels in neuronal membranes, disrupting neuronal homeostasis [82–87]. Normally, an influx of Ca^{2+} ions is strictly controlled, evoking the release of neurotransmitters like glutamate from presynaptic terminals and triggering downstream signals that regulate cellular processes including synaptogenesis, synaptic transmission, synaptic plasticity, neuronal development, and survival. However, in AD, Ca^{2+} flux is disrupted as a result of increased oxidative stress and disrupted energy metabolism, which affects the glutamate receptor, glucose transporters, and ion-motive ATPases' normal function [88]. In the hippocampus and cortex region in the brain, for instance, accumulated Aβ has been found to elevate the cellular Ca^{2+} ion level by plasma membrane L-type Ca^{2+} channels and Na^+/K^+- ATPase activity causing extreme excitatory responses, i.e., glutamate excitotoxicity and neuronal mortality [89]. Presenilin-1, the catalytic subunit of γ secretase, is also identified to be responsible for leaking Ca^{2+} ions from the endoplasmic reticulum to the cytoplasm via Ca^{2+} leak channels, increasing the cellular burden of Ca^{2+} ion in the AD brain [90]. Additionally, new research has revealed that transient receptor potential (TRP) channels impair Ca^{2+} homeostasis in Alzheimer's disease. Thus, elevated intracellular Ca^{2+} ion alters amyloid-β precursor protein (AβPP) processing and influences various downstream pathways, including tau metabolism, housekeeping gene suppression, and autophagic function loss, worsening the symptoms of AD [91]. K^+ channel abnormalities have also been identified in AD patients. Because the K^+ channel is essential for the formation of action potentials and the maintenance of the resting potential, any blockage causes poor neurotransmission and neuronal injury. Furthermore, accumulating Aβ has been found in hippocampus neurons to suppress voltage-dependent fast-inactivating K^+ currents [92]. Moreover, Kv1.3, Kv1.5, KCNN4/KCa3.1 respectively voltage-gated K^+ channels and calcium-activated K^+ channel, have been reported to induce neurodegeneration in response to neuroinflammation caused by Aβ peptides via microglial activation [93]. Similarly, the Kv3 subfamilies of K^+ channel subunits, which can rapidly repolarize the action potential, have been reported to be impaired and downregulated in AD [94]. As a result of the increased K^+ channel activity, intracellular Ca^{2+} overload occurs, leading to altered neuronal excitability and perhaps neuronal death [95]. On the plasma membrane of activated microglial cells in the hippocampus of mild AD patients, a novel intracellular chloride channel 1 (CLIC1) was recently discovered. CLIC1 channels become strongly expressed after Aβ stimulation of microglia and are responsible for the change in membrane anion permeability of the cell, resulting in neuronal death [96]. In addition to these channels, nAChR also plays a key role in the AD brain because cholinergic depletion may raise the production of Aβ and exacerbate its neurotoxicity through an alteration of the signal transduction events combined with cholinergic neurotransmission [97]. Additionally, the expressions of nAChR subtypes, are described to be highly expressed in AD-affected brain regions, thereby suggesting a role of these receptors in the AD etiopathology [98]. Tan and colleagues reviewed different calcium channel blockers dihydropyridines, benzothiazepines, and phenylalkylamines [99]. Moreover, Wiseman and Jarvik also reviewed different patents on potassium channel blockers or activators as possible therapies against AD such as 2-(phenylamino) benzimidazole, 2-amino benzimidazole derivatives, bis-benzimidazoles & related compounds, and many others [100]. Last but not least, as possible sodium channel blockers with useful property against AD, Shaikh and colleagues propose Aptiom (eslicarbazepine acetate) [101]. Unfortunately, we are still a long way from real AD therapy.

4.2 Ion channels and Parkinson's disease

After AD, PD is the most prevalent brain disease, affecting 1% of the elderly population (60–65 years). It is characterized by bradykinesia, postural instability stiffness, and resting tremor. PD pathogenesis is caused by a number of variables, including activities linked to cellular Ca^{2+} excess, mitochondrial malfunction, oxidative or metabolic stress, and, in particular, a small number of neurotoxins that render neuronal cells more susceptible to cell death [102]. For example, the ATP-sensitive potassium channel Kir6.2, which induces excitotoxicity, is abundantly expressed in dopaminergic (DA) neurons in the substantia nigra pars compacta (SNc) and has been linked to disease development [103]. Similarly, mutations in the Kir3.2 channel render it nonselective, producing an increase in the conduction of Na^+ ions as a replacement for highly selective K^+ ions, resulting in the loss of cerebellar cells and DA neurons in the SNc [55]. An important study by Sheih and colleagues also demonstrate that Kir3.1 and Kir3.2 are involved in the direct degeneration of DA neurons in the PD brain. Similarly, also voltage-gated T-type Ca^{2+} channels (TTCCs), Ca^{2+}–sensitive voltage-gated A-type K^+ channels, voltage-gated LTCCs (L-type Ca^{2+} channels), and ATP-sensitive K^+ (K-ATP) channels contribute toward basal ganglia dysfunction in SNc DA neurons thereby leading to progressive loss of neuronal firing, thus causing PD [55]. In addition to a subset of medial SNc DA neurons, K-ATP channel activation aided the transition from tonic firing to NMDAR-mediated bursting in vivo, resulting in phasic DA release. When glutamate binds to the receptor, it causes the NMDAR channel to open, allowing Ca^{2+} to flow into the cell. As a result, any changes in glutamate transmission generate dyskinesias in people with PD [104]. Recently, it was discovered that a new ion channel, the Hv1 proton channel, is expressed in human brain microglia and immune tissues and that it is required for NADPH oxidase superoxide generation during the respiratory burst in phagocytic leukocytes, which can lead to neurodegeneration such as PD [105]. Also, in this case, different studies proposed ion channel modulators against PD such as 4-amino-7-chloroquinoline, Safinamide, verapamil (phenylalkylamine), and diltiazem (benzothiazepine) [106, 107].

4.3 Ion channels and Huntington's disease

HD is a hereditary neurodegenerative disorder characterized by cognitive loss, emotional imbalance, and uncoordinated movements. It is caused by an autosomal dominant mutation in the Huntingtin (Htt) gene responsible for the expansion of CAG trinucleotide repeat >36 that leads to the synthesis of polyglutamine tract, thus mutated HTT (mHTT) protein is prone to aggregation and found to form intracellular accumulations in different cell types [108]. Using mouse models, Tong et al. studied the functional implications of ion channels in several cell types to determine the etiopathology of HD. In mHTT-expressing striatal astrocytes, altered Kir4.1 channel activity impaired extracellular K^+ homeostasis, resulting in hyperexcitability, i.e., HD motor symptoms in striatal neurons. However, the normal Kir4.1 channel is one of the most important astrocytal K^+ channels, since it is required for cell resting membrane potential and extracellular K^+ buffering in the brain [109]. Furthermore, mHTT has been shown to affect the function of high-voltage-activated (HVA) Ca^{2+} channels in HD [110]. Aside from Ca^{2+} channel malfunction, also Na^+, K^+, Cl^- ion channels have demonstrated lower expression in HD animal models in multiple studies [111]. In striatal neurons of HD transgenic mice models, other researchers discovered the reduced expression of K^+ channel subunits. Furthermore, in the R6/2 HD mouse model, expression of the muscular ClC-1 chloride channel is significantly reduced; thus, functional alteration of these channels disrupts ion homeostasis in cortical pyramidal neurons, affecting neurotransmitter release, synaptic integration, and genetic expression, all of which contribute to cortical dysfunction in HD [112, 113]. For HD, different calcium channel modulator has been proposed, such as 6-amino-4-(4-phenoxyphenethyl-amino)quinazoline (EVP4593), Inositol 1,4,5-Trisphosphate (Ip_3-sponge), Brilliant Blue G, and others, but the way is still long [114, 115].

4.4 Ion channels in multiple sclerosis and amyotrophic lateral sclerosis

MS is an immune-mediated central nervous system degenerative condition characterized by progressive demyelination in patches throughout the brain and spinal cord. Loss of coordination, muscle weakness, visual, and lingual problems are common signs of this condition, which affects young people in industrialized cultures the most. The presence of macrophages, T lymphocytes, microglia, and dendritic cells has been associated with inflammatory neuronal injury [116]. Infiltrating lymphocytes and macrophages harm neurons largely by direct cell contact or toxicity mediators such as glutamate or nitric oxide, as well as indirectly through the loss of oligodendrocytes and myelin sheath. Apart from inflammatory mediators, redistribution of voltage/ligand-gated ion channels and transporters has been linked to intracellular calcium excess, mitochondrial dysfunction, changes in electrical activity, and neuronal death [117]. Further, alterations in the expression pattern of Nav1.2, Nav1.5, Nav1.6, and Nav1.8, specific voltage-gated Na^+ channel isoforms have been reported in MS, and their overworking is involved in axonal deterioration followed by cerebellar dysfunction [118]. Furthermore, Nav channels cause a Na^+ influx into axons, which raises the amount of intra-axonal Ca^{2+} ions and interferes with axon myelination, resulting in MS pathogenicity. Different calcium and potassium channel isoforms were also shown to be increased, interfering with conduction in demyelinating axons [119]. ALS is a fatal chronic motor neurodegenerative disease marked by significant motor neuron loss in the motor cortex, brain stem, and spinal cord. Patients develop progressive muscle weakening, fasciculation, and atrophy, which leads to a loss of voluntary movement [120]. However, the specific etiology of ALS remains unknown, however, animal models are being used in research to find a feasible reason. Previous research revealed that the contraction of mammalian denervated muscle fibers is caused by spontaneous activation of the voltage-gated Na^+ channel [121]. Furthermore, in human sporadic ALS, a significant drop in potassium channel expression has been found [122]. Axonal hyperexcitability is caused by continuous Na^+ ion conduction followed by a rapid reduction in K^+ ion conductance, resulting in ALS symptoms [123]. Furthermore, in ALS, motor neurons that innervate tongue muscles are prone to degeneration, which has been related to VGCC expression differences. In ALS patients and animal models, other investigations have found immunoreactivity with several calcium ion channels [124]. Israelson et al. investigated the role of mitochondrial channelopathy during ALS and discovered that mutant superoxide dismutase 1 (SOD1) inhibits the mitochondrial voltage-dependent anion channel-1 and induces mitochondrial-dependent apoptosis, resulting in lethal paralysis in ALS. However, further study is being conducted to determine the specific mechanism responsible for its etiology [43, 125]. There is a lack of data to address the review question on the efficacy of Na^+ channel blockers for people with MS [126]. The K^+ channel blocker Fampridine-SR is an authorized MS therapy adjunct that has been demonstrated to help with ambulation, tiredness, and endurance [127]. Silva et al. demonstrated the efficacy of Ca^{2+} channel blocker CTK 01512-2 in mouse models of MS comparing it with Ziconotide. They found a significant improvement in neuroinflammatory event MS-related [128].

5. Conclusion

Ion channel dysfunction is steadily becoming connected to neurological disorders, making it an intriguing subject of neuroscience research. It has been linked to memory loss, movement issues, and neuromuscular anomalies in a number of neurological diseases. Since they originate in response to genetic defects in channel coding proteins that disturb the ionic equilibrium in the brain, the majority of these illnesses are classified as neurological channelopathies. Aging is a complex and multidimensional biological process that affects all organ systems. In the core section of them, cellular malfunction and senescent cell accumulation are common. Various aspects of brain aging have been discovered at the molecular, cellular, and tissue levels, according to research in the fields of aging and neurobiology. Aging is the leading risk factor for a broad range of neurodegenerative disorders. According to breakthroughs in the treatment and prevention of some tough diseases such as cardiovascular disease and malignancies, which have enabled more people to survive past the age of 70, aging brain disorders have lately become the leading cause of disability and death. After providing an overview of recent developments in brain aging, the current review describes it as the result of decreased neurogenesis and synaptic plasticity, as well as altered neurochemical and signaling pathways, such as impaired protein processing, glial cell activation, impaired mitochondrial function, increased oxidative stress, and neuroinflammation. Furthermore, the hippocampus and neocortex are the principal susceptible sections, with varying degrees of molecular and cellular abnormalities in their sub-centers as a result of aging. Although each of these age-related alterations is present during normal aging, their combined influence, when combined with genetic background and environmental variables, may trigger the cytotoxic activation cycle. Transcript factors, proteins, and cell-environmental variables including redox potential are all connected to these alterations. However, the crucial component that governs the entire activity is unclear. One of the first priorities would be to figure out how redox capability influences gene transcription and promotes metabolic responses as the brain matures. Our understanding of brain aging is still in its early stages. More research is needed to discover effective therapy approaches and drugs to combat brain aging. Furthermore, non-pharmacological techniques such as lifestyle adjustments, physical exercise, and calorie restriction, which promote the brain's physiological processes while reducing ROS formation and inflammation, may help to promote healthy aging. Understanding the processes that underpin the hallmarks is crucial for developing future therapeutics to slow or even reverse the aging process in the brain. The primary objective of neurobiology and brain aging research should be to discover methods and techniques for supporting healthy brain aging in all people. The functional activities of ion channels connected to the onset of numerous chronic neurological diseases have been determined. NDDs are accompanied with inflammations, neurotoxic protein accumulations, physiological stress, and mitochondrial dysfunctions, according to experimental findings. These abnormal alterations cause disruptions in normal physiological processes and brain homeostasis, which leads to illness development. The pathogenesis of AD, PD, HD, MS, and ALS has been further clarified in terms of faulty ion channels. Until today many natural compounds or synthetics compounds are identified as a modulator of ion channels (for a very extensive review refer to [43, 129]). Furthermore, channel modulators have been discovered to be important in correcting the chronic consequences of abnormal ion channels. Furthermore, understanding their regulation mechanisms in neurodegeneration might lead to the development of newer, more effective treatment techniques.

Conflict of interest

The authors declare no conflict of interest.

Author details

Marika Cordaro[1], Salvatore Cuzzocrea[2] and Rosanna Di Paola[2]*

1 Department of Biomedical and Dental Sciences and Morphofunctional Imaging, University of Messina, Messina, Italy

2 Department of Chemical, Biological, Pharmaceutical and Environmental Sciences, University of Messina, Messina, Italy

*Address all correspondence to: dipaolar@unime.it

References

[1] Tosato M et al. The aging process and potential interventions to extend life expectancy. Clinical Interventions in Aging. 2007;**2**(3):401-412

[2] Zia A et al. Molecular and cellular pathways contributing to brain aging. Behavioral and Brain Functions. 2021;**17**(1):6

[3] Mariani E et al. Oxidative stress in brain aging, neurodegenerative and vascular diseases: An overview. Journal of Chromatography. B, Analytical Technologies in the Biomedical and Life Sciences. 2005;**827**(1):65-75

[4] Strickland M et al. Relationships between ion channels, mitochondrial functions and inflammation in human aging. Frontiers in Physiology. 2019;**10**:158

[5] Patel R, Sesti F. Oxidation of ion channels in the aging nervous system. Brain Research. 2016;**1639**:174-185

[6] Exercise for female cardiac patients? Several factors play a role in continued commitment. American Journal of Nursing. 2003;**103**(10):17-17

[7] Floyd RA, Hensley K. Oxidative stress in brain aging. Implications for therapeutics of neurodegenerative diseases. Neurobiology of Aging. 2002;**23**(5):795-807

[8] Huffman K. The developing, aging neocortex: How genetics and epigenetics influence early developmental patterning and age-related change. Frontiers in Genetics. 2012;**3**:212

[9] Lemoine M. The evolution of the hallmarks of aging. Frontiers in Genetics. 2021;**12**:693071

[10] Peng K et al. Mitochondrial ATP-sensitive potassium channel regulates mitochondrial dynamics to participate in neurodegeneration of Parkinson's disease. Biochimica et Biophysica Acta (BBA)-Molecular Basis of Disease. 2018;**1864**(4 Pt A):1086-1103

[11] Bokov A, Chaudhuri A, Richardson A. The role of oxidative damage and stress in aging. Mechanisms of Ageing and Development. 2004;**125**(10-11):811-826

[12] Cini M, Moretti A. Studies on lipid peroxidation and protein oxidation in the aging brain. Neurobiology of Aging. 1995;**16**(1):53-57

[13] Halliwell B. Role of free radicals in the neurodegenerative diseases: Therapeutic implications for antioxidant treatment. Drugs & Aging. 2001;**18**(9):685-716

[14] Park L et al. Nox2-derived reactive oxygen species mediate neurovascular dysregulation in the aging mouse brain. Journal of Cerebral Blood Flow and Metabolism. 2007;**27**(12):1908-1918

[15] Gorrini C, Harris IS, Mak TW. Modulation of oxidative stress as an anticancer strategy. Nature Reviews. Drug Discovery. 2013;**12**(12):931-947

[16] Calabrese V et al. Disruption of thiol homeostasis and nitrosative stress in the cerebrospinal fluid of patients with active multiple sclerosis: Evidence for a protective role of acetylcarnitine. Neurochemical Research. 2003;**28**(9):1321-1328

[17] Delves PJ, Roitt IM. The immune system. Second of two parts. The New England Journal of Medicine. 2000;**343**(2):108-117

[18] Gruver AL, Hudson LL, Sempowski GD. Immunosenescence of ageing. The Journal of Pathology. 2007;**211**(2):144-156

[19] Franceschi C et al. An evolutionary perspective on immunosenescence. Annals of the New York Academy of Sciences. 2000;**908**:244-254

[20] Salminen A, Kauppinen A, Kaarniranta K. Emerging role of NF-kappaB signaling in the induction of senescence-associated secretory phenotype (SASP). Cellular Signalling. 2012;**24**(4):835-845

[21] Yaffe K et al. Inflammatory markers and cognition in well-functioning African-American and white elders. Neurology. 2003;**61**(1):76-80

[22] Mosher KI, Wyss-Coray T. Microglial dysfunction in brain aging and Alzheimer's disease. Biochemical Pharmacology. 2014;**88**(4):594-604

[23] von Bernhardi R, Eugenin-von Bernhardi L, Eugenin J. Microglial cell dysregulation in brain aging and neurodegeneration. Frontiers in Aging Neuroscience. 2015;**7**:124

[24] Kim SU, de Vellis J. Microglia in health and disease. Journal of Neuroscience Research. 2005;**81**(3):302-313

[25] Block ML, Zecca L, Hong JS. Microglia-mediated neurotoxicity: Uncovering the molecular mechanisms. Nature Reviews. Neuroscience. 2007;**8**(1):57-69

[26] Liu X et al. Age-dependent neuroinflammatory responses and deficits in long-term potentiation in the hippocampus during systemic inflammation. Neuroscience. 2012;**216**:133-142

[27] Raj DD et al. Priming of microglia in a DNA-repair deficient model of accelerated aging. Neurobiology of Aging. 2014;**35**(9):2147-2160

[28] Snow WM et al. Roles for NF-kappaB and gene targets of NF-kappaB in synaptic plasticity, memory, and navigation. Molecular Neurobiology. 2014;**49**(2):757-770

[29] Camandola S, Mattson MP. Brain metabolism in health, aging, and neurodegeneration. The EMBO Journal. 2017;**36**(11):1474-1492

[30] Rose CR et al. Astroglial glutamate signaling and uptake in the Hippocampus. Frontiers in Molecular Neuroscience. 2017;**10**:451

[31] Barger SW et al. Tumor necrosis factors alpha and beta protect neurons against amyloid beta-peptide toxicity: Evidence for involvement of a kappa B-binding factor and attenuation of peroxide and Ca2+ accumulation. Proceedings of the National Academy of Sciences of the United States of America. 1995;**92**(20):9328-9332

[32] Kassed CA et al. Mice expressing human mutant presenilin-1 exhibit decreased activation of NF-kappaB p50 in hippocampal neurons after injury. Brain Research. Molecular Brain Research. 2003;**110**(1):152-157

[33] Kassed CA et al. Lack of NF-kappaB p50 exacerbates degeneration of hippocampal neurons after chemical exposure and impairs learning. Experimental Neurology. 2002;**176**(2):277-288

[34] Pennypacker KR et al. NF-kappaB p50 is increased in neurons surviving hippocampal injury. Experimental Neurology. 2001;**172**(2):307-319

[35] Bruccoleri A, Pennypacker KR, Harry GJ. Effect of dexamethasone on elevated cytokine mRNA levels in chemical-induced hippocampal injury. Journal of Neuroscience Research. 1999;**57**(6):916-926

[36] Pennypacker K, Fischer I, Levitt P. Early in vitro genesis and differentiation of axons and dendrites by hippocampal neurons analyzed quantitatively with neurofilament-H and microtubule-associated protein 2 antibodies. Experimental Neurology. 1991;**111**(1):25-35

[37] Wang X et al. Cytosolic prion protein induces apoptosis in human neuronal cell SH-SY5Y via mitochondrial disruption pathway. BMB Reports. 2009;**42**(7):444-449

[38] Bryant L et al. Spinal ceramide and neuronal apoptosis in morphine antinociceptive tolerance. Neuroscience Letters. 2009;**463**(1):49-53

[39] Gao HL et al. Zinc deficiency reduces neurogenesis accompanied by neuronal apoptosis through caspase-dependent and -independent signaling pathways. Neurotoxicity Research. 2009;**16**(4):416-425

[40] Wu JS et al. Ligand-activated peroxisome proliferator-activated receptor-gamma protects against ischemic cerebral infarction and neuronal apoptosis by 14-3-3 epsilon upregulation. Circulation. 2009;**119**(8):1124-1134

[41] Li J et al. Protection of PMS777, a new AChE inhibitor with PAF antagonism, against amyloid-beta-induced neuronal apoptosis and neuroinflammation. Cellular and Molecular Neurobiology. 2009;**29**(4):589-595

[42] Chen L et al. Hydrogen peroxide-induced neuronal apoptosis is associated with inhibition of protein phosphatase 2A and 5, leading to activation of MAPK pathway. The International Journal of Biochemistry & Cell Biology. 2009;**41**(6):1284-1295

[43] Kumar P et al. Ion channels in neurological disorders. Advances in Protein Chemistry and Structural Biology. 2016;**103**:97-136

[44] Cooper EC, Jan LY. Ion channel genes and human neurological disease: Recent progress, prospects, and challenges. Proceedings of the National Academy of Sciences of the United States of America. 1999;**96**(9):4759-4766

[45] Lodish H et al. Molecular Cell Biology. New York City, United States: Macmillan; 2008

[46] Barnett MW, Larkman PM. The action potential. Practical Neurology. 2007;**7**(3):192-197

[47] Strange K. Cellular volume homeostasis. Advances in Physiology Education. 2004;**28**(1-4):155-159

[48] Hoppa MB et al. Control and plasticity of the presynaptic action potential waveform at small CNS nerve terminals. Neuron. 2014;**84**(4):778-789

[49] Rowan MJ et al. Synapse-level determination of action potential duration by K(+) channel clustering in axons. Neuron. 2016;**91**(2):370-383

[50] Charpentier M et al. Nuclear-localized cyclic nucleotide-gated channels mediate symbiotic calcium oscillations. Science. 2016;**352**(6289):1102-1105

[51] Raffaello A et al. Calcium at the center of cell signaling: Interplay between

endoplasmic reticulum, mitochondria, and lysosomes. Trends in Biochemical Sciences. 2016;**41**(12):1035-1049

[52] Kurachi Y, North A. Ion channels: Their structure, function and control–an overview. The Journal of Physiology. 2004;**554**(2):245-247

[53] Santulli G, Marks AR. Essential roles of intracellular calcium release channels in muscle, brain, metabolism, and aging. Current Molecular Pharmacology. 2015;**8**(2):206-222

[54] Rao V, Kaja S, Gentile S. Ion Channels in Aging and Aging-Related Diseases. In: Shiomi N, editor. Molecular Mechanisms of the Aging Process and Rejuvenation [Internet]. London: IntechOpen; 2016 [cited 2022 Mar 23]. Available from: https://www.intechopen.com/chapters/51760. DOI: 10.5772/63951

[55] Shieh CC et al. Potassium channels: Molecular defects, diseases, and therapeutic opportunities. Pharmacological Reviews. 2000;**52**(4):557-594

[56] Tian C et al. Potassium channels: Structures, diseases, and modulators. Chemical Biology & Drug Design. 2014;**83**(1):1-26

[57] Colom LV et al. Role of potassium channels in amyloid-induced cell death. Journal of Neurochemistry. 1998;**70**(5):1925-1934

[58] Annunziato L et al. Modulation of ion channels by reactive oxygen and nitrogen species: A pathophysiological role in brain aging? Neurobiology of Aging. 2002;**23**(5):819-834

[59] Yu SP, Canzoniero LM, Choi DW. Ion homeostasis and apoptosis. Current Opinion in Cell Biology. 2001;**13**(4):405-411

[60] Yu SP et al. Role of the outward delayed rectifier K+ current in ceramide-induced caspase activation and apoptosis in cultured cortical neurons. Journal of Neurochemistry. 1999;**73**(3):933-941

[61] Neher E, Sakaba T. Multiple roles of calcium ions in the regulation of neurotransmitter release. Neuron. 2008;**59**(6):861-872

[62] Nikoletopoulou V, Tavernarakis N. Calcium homeostasis in aging neurons. Frontiers in Genetics. 2012;**3**:200

[63] Toescu EC. Altered calcium homeostasis in old neurons. In: Riddle DR, editor. Brain Aging: Models, Methods, and Mechanisms. Boca Raton (FL): Taylor & Francis Group; 2007

[64] Griffith WH et al. Modification of ion channels and calcium homeostasis of basal forebrain neurons during aging. Behavioural Brain Research. 2000;**115**(2):219-233

[65] Gibson GE, Peterson C. Calcium and the aging nervous system. Neurobiology of Aging. 1987;**8**(4):329-343

[66] Disterhoft JF, Moyer JR Jr, Thompson LT. The calcium rationale in aging and Alzheimer's disease. Evidence from an animal model of normal aging. Annals of the New York Academy of Sciences. 1994;**747**:382-406

[67] Khachaturian ZS. Calcium, membranes, aging, and Alzheimer's disease. Introduction and overview. Annals of the New York Academy of Sciences. 1989;**568**:1-4

[68] Gant JC et al. Aging-related calcium dysregulation in rat entorhinal neurons homologous with the human entorhinal neurons in which Alzheimer's disease neurofibrillary tangles first appear. Journal of Alzheimer's Disease. 2018;**66**(4):1371-1378

[69] Michaelis ML et al. Decreased plasma membrane calcium transport activity in aging brain. Life Sciences. 1996;**59**(5-6):405-412

[70] Thibault O, Gant JC, Landfield PW. Expansion of the calcium hypothesis of brain aging and Alzheimer's disease: Minding the store. Aging Cell. 2007;**6**(3):307-317

[71] Malenka RC et al. The impact of postsynaptic calcium on synaptic transmission—Its role in long-term potentiation. Trends in Neurosciences. 1989;**12**(11):444-450

[72] Bliss TV, Collingridge GL. A synaptic model of memory: Long-term potentiation in the hippocampus. Nature. 1993;**361**(6407):31-39

[73] Celsi F et al. Mitochondria, calcium and cell death: A deadly triad in neurodegeneration. Biochimica et Biophysica Acta. 2009;**1787**(5):335-344

[74] Catterall WA. From ionic currents to molecular mechanisms: The structure and function of voltage-gated sodium channels. Neuron. 2000;**26**(1):13-25

[75] Catterall WA. Voltage-gated sodium channels at 60: Structure, function and pathophysiology. The Journal of Physiology. 2012;**590**(11):2577-2589

[76] Caring for a dog with osteoarthritis. The Veterinary Record. 2018;**182**(15):440

[77] Desaphy JF et al. Modification by ageing of the tetrodotoxin-sensitive sodium channels in rat skeletal muscle fibres. Biochimica et Biophysica Acta. 1998;**1373**(1):37-46

[78] Veal EA, Day AM, Morgan BA. Hydrogen peroxide sensing and signaling. Molecular Cell. 2007;**26**(1):1-14

[79] Simms BA, Zamponi GW. Neuronal voltage-gated calcium channels: Structure, function, and dysfunction. Neuron. 2014;**82**(1):24-45

[80] Zhao Y, Zhao B. Oxidative stress and the pathogenesis of Alzheimer's disease. Oxidative Medicine and Cellular Longevity. 2013;**2013**:316523

[81] Hardy J. Alzheimer's disease: The amyloid cascade hypothesis: An update and reappraisal. Journal of Alzheimer's Disease. 2006;**9**(3 Suppl):151-153

[82] Neelands TR, King AP, Macdonald RL. Functional expression of L-, N-, P/Q-, and R-type calcium channels in the human NT2-N cell line. Journal of Neurophysiology. 2000;**84**(6):2933-2944

[83] Nimmrich V, Eckert A. Calcium channel blockers and dementia. British Journal of Pharmacology. 2013;**169**(6):1203-1210

[84] Kim S, Rhim H. Effects of amyloid-beta peptides on voltage-gated L-type Ca(V)1.2 and Ca(V)1.3 Ca(2+) channels. Molecules and Cells. 2011;**32**(3):289-294

[85] Danysz W, Parsons CG. Alzheimer' disease, beta-amyloid, glutamate, NMDA receptors and memantine—Searching for the connections. British Journal of Pharmacology. 2012;**167**(2):324-352

[86] Oules B et al. Ryanodine receptor blockade reduces amyloid-beta load and memory impairments in Tg2576 mouse model of Alzheimer disease. The Journal of Neuroscience. 2012;**32**(34):11820-11834

[87] Parri RH, Dineley TK. Nicotinic acetylcholine receptor interaction with beta-amyloid: Molecular, cellular, and physiological consequences. Current Alzheimer Research. 2010;**7**(1):27-39

[88] Itkin A et al. Calcium ions promote formation of amyloid beta-peptide (1-40) oligomers causally implicated in neuronal toxicity of Alzheimer's disease. PLoS One. 2011;**6**(3):e18250

[89] Camandola S, Mattson MP. Aberrant subcellular neuronal calcium regulation in aging and Alzheimer's disease. Biochimica et Biophysica Acta. 2011;**1813**(5):965-973

[90] Tu H et al. Presenilins form ER Ca2+ leak channels, a function disrupted by familial Alzheimer's disease-linked mutations. Cell. 2006;**126**(5):981-993

[91] Yamamoto S et al. Transient receptor potential channels in Alzheimer's disease. Biochimica et Biophysica Acta. 2007;**1772**(8):958-967

[92] Poulopoulou C et al. Aberrant modulation of a delayed rectifier potassium channel by glutamate in Alzheimer's disease. Neurobiology of Disease. 2010;**37**(2):339-348

[93] Kaushal V et al. The Ca2+−activated K+ channel KCNN4/KCa3.1 contributes to microglia activation and nitric oxide-dependent neurodegeneration. The Journal of Neuroscience. 2007;**27**(1):234-244

[94] Bhullar KS, Rupasinghe HP. Polyphenols: Multipotent therapeutic agents in neurodegenerative diseases. Oxidative Medicine and Cellular Longevity. 2013;**2013**:891748

[95] Mocali A et al. Altered proteolysis in fibroblasts of Alzheimer patients with predictive implications for subjects at risk of disease. International Journal of Alzheimer's Disease. 2014;**2014**:520152

[96] Novarino G et al. Involvement of the intracellular ion channel CLIC1 in microglia-mediated beta-amyloid-induced neurotoxicity. The Journal of Neuroscience. 2004;**24**(23):5322-5330

[97] Buckingham SD et al. Nicotinic acetylcholine receptor signalling: Roles in Alzheimer's disease and amyloid neuroprotection. Pharmacological Reviews. 2009;**61**(1):39-61

[98] Nery AA et al. Rescue of amyloid-Beta-induced inhibition of nicotinic acetylcholine receptors by a peptide homologous to the nicotine binding domain of the alpha 7 subtype. PLoS One. 2013;**8**(7):e67194

[99] Tan Y, Deng Y, Qing H. Calcium channel blockers and Alzheimer's disease. Neural Regeneration Research. 2012;**7**(2):137-140

[100] Judge SI et al. Potassium channel blockers and openers as CNS neurologic therapeutic agents. Recent Patents on CNS Drug Discovery. 2007;**2**(3):200-228

[101] Shaikh S et al. Aptiom (eslicarbazepine acetate) as a dual inhibitor of beta-secretase and voltage-gated sodium channel: Advancement in Alzheimer's disease-epilepsy linkage via an enzoinformatics study. CNS & Neurological Disorders Drug Targets. 2014;**13**(7):1258-1262

[102] Massano J, Bhatia KP. Clinical approach to Parkinson's disease: Features, diagnosis, and principles of management. Cold Spring Harbor Perspectives in Medicine. 2012;**2**(6):a008870

[103] Gong XG et al. Da-bu-yin-wan and qian-zheng-san to neuroprotect the mouse model of Parkinson's disease. Evidence-based Complementary and Alternative Medicine. 2014;**2014**:729195

[104] Schiemann J et al. K-ATP channels in dopamine substantia nigra neurons

control bursting and novelty-induced exploration. Nature Neuroscience. 2012;**15**(9):1272-1280

[105] Wu LJ et al. The voltage-gated proton channel Hv1 enhances brain damage from ischemic stroke. Nature Neuroscience. 2012;**15**(4):565-573

[106] Kim CH, Leblanc P, Kim KS. 4-amino-7-chloroquinoline derivatives for treating Parkinson's disease: Implications for drug discovery. Expert Opinion on Drug Discovery. 2016;**11**(4):337-341

[107] Ritz B et al. L-type calcium channel blockers and Parkinson disease in Denmark. Annals of Neurology. 2010;**67**(5):600-606

[108] Labbadia J, Morimoto RI. Huntington's disease: Underlying molecular mechanisms and emerging concepts. Trends in Biochemical Sciences. 2013;**38**(8):378-385

[109] Tong X et al. Astrocyte Kir4.1 ion channel deficits contribute to neuronal dysfunction in Huntington's disease model mice. Nature Neuroscience. 2014;**17**(5):694-703

[110] Miller BR, Bezprozvanny I. Corticostriatal circuit dysfunction in Huntington's disease: Intersection of glutamate, dopamine and calcium. Future Neurology. 2010;**5**(5):735-756

[111] Oyama F et al. Sodium channel beta4 subunit: Down-regulation and possible involvement in neuritic degeneration in Huntington's disease transgenic mice. Journal of Neurochemistry. 2006;**98**(2):518-529

[112] Ariano MA et al. Striatal potassium channel dysfunction in Huntington's disease transgenic mice. Journal of Neurophysiology. 2005;**93**(5):2565-2574

[113] Waters CW et al. Huntington disease skeletal muscle is hyperexcitable owing to chloride and potassium channel dysfunction. Proceedings of the National Academy of Sciences of the United States of America. 2013;**110**(22):9160-9165

[114] Czeredys M. Dysregulation of neuronal calcium signaling via store-operated channels in Huntington's disease. Frontiers in Cell and Development Biology. 2020;**8**:611735

[115] Jo S, Bean BP. Inhibition of neuronal voltage-gated sodium channels by brilliant blue G. Molecular Pharmacology. 2011;**80**(2):247-257

[116] Fitzner D, Simons M. Chronic progressive multiple sclerosis—Pathogenesis of neurodegeneration and therapeutic strategies. Current Neuropharmacology. 2010;**8**(3):305-315

[117] Meuth SG et al. Multiple sclerosis—A channelopathy? Targeting ion channels and transporters in inflammatory neurodegeneration. Nervenarzt. 2009;**80**(4):422-429

[118] Waxman SG. Axonal conduction and injury in multiple sclerosis: The role of sodium channels. Nature Reviews. Neuroscience. 2006;**7**(12):932-941

[119] Judge SI et al. Voltage-gated potassium channels in multiple sclerosis: Overview and new implications for treatment of central nervous system inflammation and degeneration. Journal of Rehabilitation Research and Development. 2006;**43**(1):111-122

[120] Guatteo E et al. Altered calcium homeostasis in motor neurons following AMPA receptor but not voltage-dependent calcium channels' activation in a genetic model of amyotrophic lateral sclerosis. Neurobiology of Disease. 2007;**28**(1):90-100

[121] Pietrobon D. Calcium channels and channelopathies of the central nervous system. Molecular Neurobiology. 2002;**25**(1):31-50

[122] Shibuya K et al. Markedly reduced axonal potassium channel expression in human sporadic amyotrophic lateral sclerosis: An immunohistochemical study. Experimental Neurology. 2011;**232**(2):149-153

[123] Kanai K et al. Altered axonal excitability properties in amyotrophic lateral sclerosis: Impaired potassium channel function related to disease stage. Brain. 2006;**129**(Pt 4):953-962

[124] Gonzalez LE et al. Amyotrophic lateral sclerosis-immunoglobulins selectively interact with neuromuscular junctions expressing P/Q-type calcium channels. Journal of Neurochemistry. 2011;**119**(4):826-838

[125] Israelson A et al. Misfolded mutant SOD1 directly inhibits VDAC1 conductance in a mouse model of inherited ALS. Neuron. 2010;**67**(4):575-587

[126] Yang C et al. Sodium channel blockers for neuroprotection in multiple sclerosis. Cochrane Database of Systematic Reviews. 2015;**10**:CD010422

[127] Morrow SA, Rosehart H, Johnson AM. The effect of Fampridine-SR on cognitive fatigue in a randomized double-blind crossover trial in patients with MS. Multiple Sclerosis and Related Disorders. 2017;**11**:4-9

[128] Silva RBM et al. Beneficial effects of the Calcium Channel blocker CTK 01512-2 in a mouse model of multiple sclerosis. Molecular Neurobiology. 2018;**55**(12):9307-9327

[129] Alexander SP et al. The concise guide to pharmacology 2021/22: Ion channels. British Journal of Pharmacology. 2021;**178**:S157-S245

Chapter 2

Mitochondrial Channels and Their Role in Cardioprotection

Keerti Mishra and Min Luo

Abstract

Mitochondria play a pivotal role in cardioprotection. The major cardioprotective mechanism is ischemic preconditioning (IpreC), through which short periods of ischemia protect a subsequent prolonged acute ischemic episode. Mitochondria channels, particularly the potassium channels (mitoK) such as ATP-dependent and calcium-activated potassium channels, have been suggested as trigger or end effectors in IpreC. Activators of mitoK are promising therapeutic agents for the treatment of the myocardial injury due to ischemic episodes. In this chapter, we are summarizing our current knowledge on the physiology function of different mitochondrial channels with a focus on the potassium channels and their mechanism in cardioprotection. Furthermore, the currently under development therapy by targeting the mitochondrial channels for the treatment of heart failure are also discussed.

Keywords: cardioprotection, ischemic preconditioning (IpreC), ischemic postconditioning (IpreC), oxidative phosphorylation, reactive oxygen species (ROS), cell death, mitochondrial permeability transition pore, mitochondrial potassium channels, ischemia, reperfusion, heart failure

1. Introduction

Heart failure is a major public health issue that is still having a poor prognosis despite all the advancements in scientific research and technologies [1]. The approaches for the drug development of heart disease are majorly relying on the pathophysiology of the cellular mechanisms and inter and intracellular channels in the failing heart. Heart being an organ of extensively high energy demand and mitochondria being the powerhouse of the eukaryotic organisms, they are meant to be closely connected. Any change in mitochondrial function inevitably affects the health of the heart irrespective of the etiology. Recent advances in the field indicate that besides having a compromised powerhouse, mitochondrial malfunctioning accompanies certain pathogenic mechanisms leading to heart failure [2, 3]. Current therapies like ischemic pre- and postconditioning provide symptomatic benefit but do not address the abnormalities at a molecular level. Since the mitochondria play an important role in the pathophysiology of a failing heart, understanding its mechanism can potentially improve the approaches for the therapies for direct improvement of cardiac functions. Among the abnormalities shown by the mitochondria, ruptured electron transport chain, excessive formation of reactive oxygen species (ROS), perturbed ion homeostasis are the basic concerns [4]. An important and potential substrate for therapeutics in heart failure is mitochondrial channels [5]. In this chapter, we intend to discuss the available information about the mitochondrial channel with regards to its pathophysiological effects on heart health and their responses to the ischemic conditioning alongside the available agonist for the mitochondrial channel.

2. Mitochondria and its functions specific to heart cells

Due to high energy demand, the number of mitochondria in the heart cells is excessively high, with a daily production of approximately 65 kg ATP through oxidative phosphorylation [6]. In the neonatal cardiac myocytes, the mitochondria are highly motile in the cytosol generating energy through glycolysis and glucose metabolism. Whereas, in an adult myocyte the mitochondria have reduced motility, and energy generation occurs from the metabolism of fatty acid [7].

Mitochondria are known to arise billions of years ago through the engulfment of alpha proteobacteria by the precursors of modern eukaryotic cells and it evolved to become an essential multifunctional organelle [8]. Mitochondria are made up of an outer comparatively permeable and inner highly folded relatively impermeable lipid bilayer. The folded inner membrane with a high surface area contains the complexes for the generation and transportation of adenosine triphosphate (ATP) through oxidative phosphorylation. In the myocardial cells, the substrates are oxidized to produce acetyl coenzyme A, which in turn drives the Krebs cycle to produce nicotinamide adenine dinucleotide hydrogen (NADH) and flavin adenine dinucleotide ($FADH_2$) in the mitochondrial matrix. The oxidation of NADH and $FADH_2$ leads to the establishment of proton motive force later fetched by F1F0 ATP synthase to convert adenosine diphosphate (ADP) and inorganic phosphate to ATP [9].

In the process of energy production, approximately 2% of the electrons flowing in the electron transport cycle are reduced to form a superoxide anion which is reduced to H_2O_2 followed by H_2O generation by antioxidant enzymes. Excessive production of these ROS is toxic to the cell, yet these natural byproducts of oxygen metabolism trigger a variety of oxygen sensing machinery including gene expression, however, the overload of ROS impairs the redox potential of the cell leading to various oxidative damages [10].

The dynamics of Ca^{2+}, an important element to trigger various enzymatic processes and a second messenger for contractile functions, is also organized by mitochondria by either transmembrane Ca^{2+} transport or ROS-mediated signaling pathways [11]. In case of increased workload rapid mitochondrial Ca^{2+} uptake is facilitated by Ca^{2+} uptake channel for elevated ATP production. The elimination of Ca^{2+} from the mitochondrial matrix however is slower and mediated either directly by Na^+/Ca^{2+} exchanger or indirectly by multiple mitochondrial K^+ channels with unknown mechanisms [12].

Alongside the role as a life-supporting system, mitochondria can also trigger programmed cell death in the required conditions. The mitochondrial permeability transition pore (MPTP) opens in response to stress and leads to loss of membrane potential, which stops ATP production and release of cytochrome C and other mitochondrial protein causing necrosis and cell apoptosis [13]. In cases of heart failure mitochondria-induced cell death is an important mechanism. Here we intend to discuss important parameters of mitochondrial dysfunction which lead to heart failure, and therapeutic approaches to circumvent the situation.

3. Mitochondrial dysfunction and heart failure

In cardiac cells the energy consumption should meet the energy production rate on a beat-by-beat basis, failing which the stored energy cannot last more than a few seconds. In a pathological remodeling the oxidative metabolism switch from fatty acid metabolism to glycolysis, which only contributes less than 5% of the total ATP demand of an adult heart [14]. On the other hand, during pathological remodeling the required energy increases due to disturbed cardiac geometry, and impaired ATP homeostasis. Studies have shown that the mitochondrial mechanisms involved in pathological remodeling in efforts to restore the energy homeostasis eventually led to a vicious cycle that drives pathological remodeling towards heart failure. The most puzzling scenario suggests that in the failing heart the ATP content is largely maintained after an initial glitch, thus whether the heart failure occurs due to energy starvation or in efforts to fight that starvation is a question yet to be addressed. Further in-depth analysis of the mitochondrial mechanisms can clarify if the efforts of maintaining the energy hemostasis are either helpful or potentially worsen the failing heart.

The catalysis of degradative oxidation of the nutrients through anaerobic dehydrogenases is facilitated by the reduction of oxidized pyridine and flavin nucleotide like NAD(P^+) and FAD. These coenzymes should be again reoxidized since they are non-replenishable and cannot permeate the cell membrane with the degradation rate. During an ischemic episode since the respiratory chain is impaired the oxidation of the above-stated substrates is also hampered, moreover, the NADH(H^+) oxidation is carried out by lactate dehydrogenase. Therefore, the anaerobic glycolysis takes over as the only pathway for ATP synthesis provided the phosphocreatinine is depleted with the onset of ischemia. Therefore, in a failing heart the oxidative metabolism switch for alternative carbon sources such as glucose which can be beneficial due to increased ATP production and oxygen uptake but when it takes over the usual fatty acid metabolism the energy production is not sufficient for an adult heart [14]. Increased glycolysis causes anaplerosis, increased lactate production, triggers the heart to go into pathological remodeling, and also inhibits branched-chain amino acid (BCAA) catabolism, and causes the accumulation of BCAA. A hyperacetylation of mitochondrial protein has also been seen as a failing heart the cause of which is not clearly understood [15].

The decrease in ATP concentration causes an immense ionic imbalance across the cell cytoplasm leading to the lowering of the pH of the cell. The inhibition of Na^+/K^+ ATPase, Na^+/H^+, Na^+/Ca^{2+} antiporters leads to an overload of Ca^{2+} inside the cells causing hypercontracture and triggering the irreversible opening of mitochondrial permeability transition pore (MPTP) [16]. The frequently converting ATP into ADP and phosphate seeps out of the cell which further contributes to reduced performance of the heart. Opening of only one pore causes frequent depolarization and triggers the opening of other pores, following which the rapid influx of small molecular weight solutes enters the mitochondrial matrix to compensate for the depolarization and causes the mitochondrial matrix to swell. The expansion of the inner mitochondrial membrane leads to the rupture of the outer membrane which releases proapoptotic proteins leading eventually to cell death. Therefore, it is believed that altering the MPTP pore opening can be helpful in the prevention of cardiac reperfusion and cell death [17].

During an ischemic episode, the release of ROS is formed under the physiological and pathological conditions within the mitochondria. In a regular respiratory chain reaction, 2–4% oxygen undergoes an univalent reaction and produces superoxide [4]. The superoxides that are formed at complex I and complex III level are rapidly transformed by metalloenzymes like superoxide dismutase into hydrogen peroxide. In the first minute, it is small but in a later stage, it increases dramatically, leading to the disruption of mitochondrial membrane potential. Therefore, the consequence of ROS formation has been linked to the opening of the MPTP channel leading to apoptosis. These episodes put together in series lead to a gradual and irreversible decline of the cell integrity.

The opening of MPTP can occur through all the factors mentioned here, such as an increase in Ca^{2+} ion, depolarization, increase in the ROS, and phosphate concentration [18]. Certain factors like a high concentration of H^+, Mg^{2+}, and ADP can counteract the MPTP opening and work as antagonists [19, 20]. On the contrary, in the condition of reperfusion, the change in the pH is recovered by the burst formation of ROS in the presence of Ca^{2+}, which creates the most favorable condition for MPTP opening even though the antagonizing effect of membrane potential recovery occurred. In isolated mitochondria, the MPTP opens at a very high Ca^{2+} concentration which is practically not possible in vivo therefore the increased Ca^{2+} alone is not responsible for MPTP opening rather can be triggered by several processes like ROS generating Ca^{2+} dependent enzyme.

4. Cardioprotection

It is believed that to reduce the damage occurring in the heart cells in a prolonged ischemic episode, the heart cells can be trained beforehand through small and regulated episodes of either cardiac ischemia or reperfusion that resulted from ATP deprivation or concentration increase of ROS and Ca^{2+} (**Figure 1**). This method has been tested in dogs [21] and higher mammals including humans [22, 23]. This process is known as ischemic preconditioning (IpreC). Similarly, ischemic postcondition

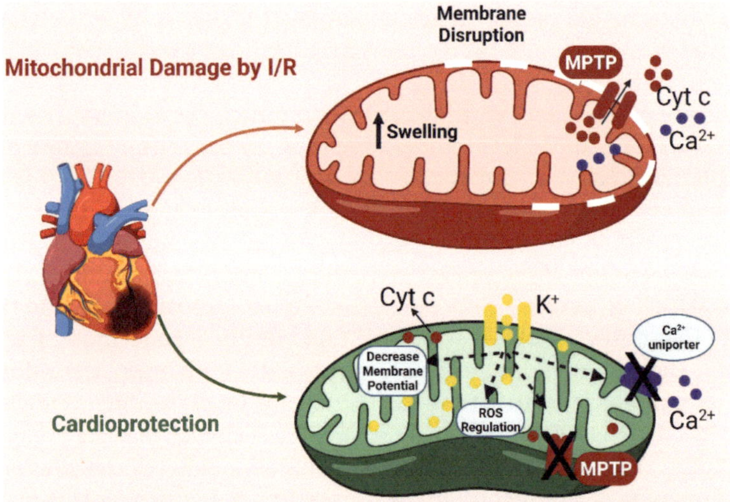

Figure 1.
The ischemic/reperfused heart mitochondria in comparison to the cardioprotected mitochondria. In cardioprotected mitochondria MPTP and MCU channels are closed and mitoK channel is opened.

(IpostC) can also be done in a brief intermittent cycle after a severe event [24]. These are performed by natural or artificial biomolecules which will be discussed later in this chapter.

The process of IpreC and IpostC usually activates protein kinase C isozymes [25] and other kinases [26] whose roles in cardioprotection are very dicey, because as ε isozyme protects the mitochondrial function by activating $ALDH_2$ aldehyde dehydrogenase which removes the lipid peroxidation products, Baines et al. showed that the translocation of ε isozyme prevents the opening of MPTP pore [27]. Whereas δ isozyme of protein kinase C increases the tissue injury by flawed perfusion of myocytes and inhibits ATP and pyruvate dehydrogenase regeneration [27]. Several mitochondrial pathways are activated in the conditioning process contributing significantly to the process of cardioprotection and therefore they are considered attractive pharmacological targets.

5. Mitochondrial channels with an integral role in cardioprotection

The multifaceted relationship of mitochondria with cell death makes it an ideal target for aiming to preserve cardiomyocytes viability. In the lack of oxygen during ischemia although the ATP synthesis cannot be restored yet can protect through decreasing ATP hydrolysis. Several self-defense mechanisms are triggered by ischemic preconditioning like the depolarization of mitochondrial matrix promotes F1F0 ATPase binding to its natural inhibiter Factor (IF) [28]. A similar effect has been shown by overexpressed the BCL-2 gene in mice hearts, to conclude that ATP hydrolysis is modulated by BCL-2 as well since the oligomycin addition did not possess any additional effect. BCL-2 is upregulated in the preconditioned heart and downregulated by ischemia and reperfusion [29]. However, the cardioprotective effect caused by preconditioning can be abolished by antisense nucleotide in a perfused rat heart [30]. Another way to prevent ATP hydrolysis is by MPTP inhibition, which presents a wide range of protective actions like maintaining Ca^{2+} homeostasis, NAD^+ depletion prevention, and preventing the release of pro-apoptotic protein [20, 31, 32]. The preconditioned heart prevents the opening of MPTP pores conferring stress-tolerant condition of the cardiomyocytes [33, 34].

In addition to protective effects posed by MPTP inhibition, numerous studies have vouched for the supporting effect of the mitochondrial potassium channel, specially mitoK_{ATP} and calcium-dependent mitoK_{Ca}. The influx of K^+ into the inner mitochondrial matrix causes depolarization, with pH increase and matrix swelling [35–37]. It is suggested that matrix swelling due to K^+ uptake compensates for the contraction of the matrix caused by increased potential difference due to lack of oxygen. The K^+ uptake and matrix swelling are suggested to increase the recovery of ATP concentration, by preventing the loss of substrate channeling which happened due to increased potential difference at the onset of reperfusion [38].

5.1 Mitochondrial permeability transition pore

A sudden increase in the permeability of the solute in the inner mitochondrial membrane (IMM) is known as the permeability transition [16]. The MPTP was first described by Haworth and Hunter in 1979, who showed that the addition of high levels of calcium to bovine myocardial mitochondria induced a nonspecific increase in permeability of the inner mitochondrial membrane [39]. Although the occurrence of permeability transition and its inhibitor as adenine dinucleotide has been known since 1950 [40]. Our understanding of mitochondrial physiology and the acceptance of the pore theory of permeability transition is greatly attributed to the study of a mitochondrial channel.

The opening of the MPTP channel causes depolarization, blocks ATP synthesis, releases Ca^{2+}, depletes pyridine nucleotide, inhibits respiration, causes matrix swelling, which subsequently leads to cytochrome C mobilization and outer mitochondrial membrane rupture which ultimately releases endonuclease G and apoptosis-inducing factor (AIF) and other proapoptotic protein to kill the cell (**Figure 1**) [41, 42]. It should be noted though, that this detrimental effect of MPTP opening occurs only when the pore opening is long-lasting [43]. Whereas the short-term opening, both in vivo and in vitro [44], is suggested to be involved in the physiological regulation of Ca^{2+} and the homeostasis of ROS [4], subsequently providing mitochondria a fast mechanism for Ca^{2+} release. According to a study performed on a mitochondria calcium uniporter (MCU), null mice had an equal I/R injury as the wildtype littermates overruling cyclophilin-D (CyPD) protection (**Figure 1**). It leads to challenging the established concept and awaits the molecular details of the myocardial reperfusion mechanism and the precise roles of the channels for answers to these contradicting observations. The potential role of MPTP opening in heart failure was recognized way before the discovery of the role of mitochondria in apoptosis.

Although the molecular nature and precise composition of MPTP remain unknown it is believed that some proteins regulate the function of MPTP like CyPD (**Figure 1**). After the observation that cyclosporin (CsA) is a potent inhibitor of MPTP opening [45, 46], Halestrap et al. demonstrated that it occurred due to an inhibition of a peptidyl-prolyl cis-trans isomerase PPIase in the matrix [47]. They further purified and demonstrated the protein to be CyPD, which is an 18 kDa matrix protein. A range of other CsA analogs and sanglifehrin A (SfA) that showed their potency in preventing MPTP opening also acted as inhibitors of PPIase of CyPD. On the other hand, the MPTP opening is also inhibited by ATP and ADP but their complexes with Mg^{2+} and other nucleotides like AMP, GTP, or GDP fail to show a similar effect, it is worth noting that none of them are transported by the adenine nucleotide translocase (ANT) [48]. Furthermore, the increased sensitivity of MPTP opening towards Ca^{2+} is attributed to the inhibition in the binding of the ATP and ADP with ANT either by depleting the matrix of adenine nucleotides or by modifying ANT by thiol [49]. Helstrap group developed a model for MPTP, where CyPD binds to ANT and they undergo conformational changes to induce pore formation under Ca^{2+} trigger, and they showed that matrix Ca^{2+} favored 'C' conformation for ANT. Several matrices facing glutamate and aspartate residues on ANT are present whose carboxyl groups might play the role of Ca^{2+} binding as there is no Ca^{2+} binding motif established on ANT [49]. Another data consistent with the model showed the coprecipitation of CyPD specifically with ANT and the bonding increases with rising oxidative stress and decreases with the introduction of CsA but not with inactive CsH analog [50, 51]. The crystal structure of bovine ANT1 [52] showed a constriction provided by 3 helices, block the channel and if these are rearranged by the change facilitated by CyPD, then an extensive conformational change might account for MPTP formation. Phosphate ion has been known as an MPTP activator and carboxyatractyloside (CAT) prevents ANT from binding the phenyl arsine oxide (PAO) column but still does not prevent MPTP activation, which suggests that PAO can have an additional MPTP activation site apart from the ANT. When CAT treated beef heart mitochondria was passed through the PAO column phosphate carrier protein (PiC) was bound to the column [53]. Pretreatment of the column with MPTP inhibitors like ubiquinone (UQo) prevents the PiC binding to the column which suggests a key role in MPTP formation. Other proteins have also been suggested to have the structural and regulatory role in MPTP formation like peripheral benzodiazepine receptor and voltage-dependent anion channel, hexokinase, creatinine kinase, BCL2 proteins, and Bcl 2 associated X (BAX) proteins may also be associated with MPTP, but which proteins eventually constitute the formation of the pore is still unknown [54, 55].

Recently, another theory of multiple pores in MPTP has been proposed. Studies have been supporting the potential roles of ANT, PiP, F1F0 ATP synthase, and CyPD to be inner membrane component but all of them has shown CsA sensitive permeability despite the genetic deletion of the responsible gene, which raises a question on the hypothesis and further investigation led to propose the multiple pore-forming mechanisms. Deletion of the C subunit of F1F0 ATP synthase showed that CsA induced MPTP synthesis showed much lower conductance as compared to wild-type MPTP [56]. This C-subunit lacking channel could be inhibited by an ANT inhibiter bongkrekic, therefore it was suggested that a classic MPTP was not formed in the knockout mitochondria. It was concluded that the MPTP formation could be enhanced through other proteins e.g. ANT in the lack of c-subunit. Another study proposed that dimer of F1F0 ATP synthase, ANT and PiC can assemble into synthasome complex, and it requires CyPD for disassembly into its components. They further suggest that ATP synthasome assembles and disassembles in high work conditions and MPTP formations respectively. Low ADP, high calcium enhancement leading to increase the membrane potential, and ROS formation trigger the disassembly of ATP synthasome leading to MPTP formation. Additional studies will be required to completely understand all the components of synthasome in generating MPTP [57].

As a result of its central role in myocardial infarction, MPTP poses itself as an obvious target for cardioprotection. A wide variety of cardioprotective protocols have been demonstrated to prevent MPTP opening during reperfusion. Certain drugs directly inhibit MPTP like CsA and SfA and their non-immunosuppressant derivatives like 4-methyl-val-CsA and D-3-MeAla-4-EtVal-CsA etc. and certain protocols that decrease oxidative stress and pH for inhibiting MPTP pore opening such as ischemic preconditioning [34] and ischemic postcondition [58], temperature preconditioning [59], Na^+/H^+ exchanger inhibiter like cariporide [60], mitochondrial ubiquinone antioxidants [61], the anesthetic propofol [62], urocortin [63], antioxidants including pyruvate [64].

The drugs that directly inhibit MPTP pose great value in protecting the heart during cardiac surgery, it has been shown that CsA improved cardiac performance following angioplasty treatment [65]. However, CsA and Sfa administration pose unwanted side effects because they interact with other cyclophilins like CypA moreover, their MPTP opening inhibition is overruled by the intensity of the pore opening stimulus [66]. This situation requires the development of new MPTP inhibitor drugs which can overcome these constraints. The development of new drugs requires structural insight into the MPTP pores.

5.2 Inner mitochondrial anion channel (IMAC)

The inner mitochondrial anion channel (IMAC) was the first mitochondrial channel to be identified using the patch-clamp method [67]. The pharmacological drug testing on the cardiomyocytes, for analysis of the mitochondrial matrix swelling, led to the discovery of its role in membrane potential perturbation. Its activity is promoted under stressed oxidizing conditions [68]. O'Rourke and co-workers proposed that the arrhythmias and electrophysiological alteration in cardiomyocytes are the results of disturbed membrane potential due to failed cellular mitochondrial network under oxidative stress [69]. The inhibition of IMAC mediated mitochondrial membrane potential oscillation with 4-chlorodiazepam showed a significant reduction and stabilization of the sarcolemmal action potential [70]. High-resolution optical action potential mapping showed that the introduction of 4-chlorodiazepam facilitates the restoration of action potential duration and prevents ventricular fibrillation. The thiol oxidants trigger the oscillation of membrane potential, glutathione, and NADH, which in turn increases the ROS concentration [71]. The inhibition of IMAC activity is triggered by the binding of 4-chlorodiazepam with benzodiazepine receptors. The inhibited IMAC preserves the membrane potential; however, the prohibited efflux of superoxide from IMAC further increases the ROS concentration [72]. The increasing ROS and decreasing glutathione concentration in the mitochondrial matrix trigger the opening of the MPTP pore, and therefore, IMAC can be considered as an instigator of MPTP opening [71].

5.3 Mitochondrial Ca^{2+} uniporter

The macromolecular structural assembly responsible for mitochondrial Ca^{2+} uptake machinery is known as the mitochondrial calcium uniporter (MCU) complex. It was initially assumed that an active uptake and passive release are required for the transport of Ca^{2+} across the inner mitochondrial membrane [73], but multiple groups showed that the uptake is energetically favored whereas efflux requires electrogenic ion-exchange [74].

Ca^{2+} uptake in mitochondria results from a single transport mechanism by a Ca^{2+} sensitive channel of mitochondria known as MCU (**Figure 1**). The molecular identification of the MCU protein complex which was closely connected to a comprehensive protein compendium MitoCarta was done in 2008 [75]. Following the establishment of a compendium Ca^{2+} sensing regulator, mitochondrial Ca^{2+} uptake 1(MICU1) was discovered in 2010 [76]. MICU1 was predicted to contain no transmembrane domain and was therefore not considered forming a pore. Later 40 kDa two transmembrane domains were identified termed MCU in 2011 [77, 78] followed by the identification of other regulatory subunits.

The concentration Ca^{2+} increases in the mitochondrial matrix during ischemia and reperfusion and this increase is proposed to activate MPTP opening [16]. Therefore, the inhibition of mitochondrial calcium uniporter is studied to reduce cell damage in I/R. Studies from MCU knockout mice in the germline [79] and MCU mutated gene [80], in both the cases the Ca^{2+} uptake was hindered leading to no MPTP opening, but neither of the situations reduced the size of cardiac infarct at the onset of I/R. In contrast, where the MCU was deleted after birth in adult hearts showed cardioprotection in an in vivo model [81]. The reason for this kind of difference is not very clear but apparently, the MCU knockout before birth could generate a more robust MPTP pore not regulated by CsA as well. Alongside MCU, the other two core structural components are mitochondrial calcium uniporter b MCUb and an ion transport component termed "essential MCU regulator" or EMRE. MCUb is closely related to MCU with 50% amino acid homology, containing two similar transmembrane domains linked with coiled-coil domain. On the other hand, EMRE is a 10 kDa protein span in the inner mitochondrial membrane that contains an aspartate-rich, highly conserved, C-terminal region, whose topology however is still unclear [82]. It was proposed by

Mootha et al. that EMRE is required for Ca^{2+} channeling activity and also helps in keeping the MICU1/MICU2 intact to the MCU complex [83].

Altering the levels of regulators of MCU complex the calcium uptake can also be regulated in mitochondria, subsequently altering the susceptibility to MPTP-induced cell death. A mitochondria Ca^{2+} uptake protein1 (MICU1) mutation causing a loss of function in a human patient is associated with ataxia, attributed to mitochondrial Ca^{2+} overload [84, 85]. In a failing heart, an increase of MICU1 and $Na^{+}/Li^{+}/Ca^{2+}$ exchanger (NCXL) has been observed to compensate for the Ca^{2+} overload [86]. MICU2 on the other hand has been observed to increase, with cardiovascular disease in both humans and mice, at the transcriptional level [87]. Mice with deletion of MICU2 showed a certain degree of diastolic dysfunction. The low ratio of MICU/MCU maintains a low threshold of calcium entry in mitochondria and the overexpression of MICU1 causes contractile dysfunction to the heart. Therefore, the rise in MICU1 and MICU2 with age and disease alters the susceptibility of calcium overload and MPTP inhibition.

In a cardiac muscle, constant rhythmic cycles of contraction are dependent on permanent uptake and release of Ca^{2+} in the cytoplasm and buffering organelles [88]. After a myocardial contraction, the removal of the Ca^{2+} from the cytoplasm is provided by Na^{+}/Ca^{2+} exchanger (NCX) in the endoplasmic reticulum, on the other hand in non-muscle cells the cytosolic Ca^{2+} signals and Ca^{2+} buffering depends on mitochondrial Ca^{2+} uptake [89]. Although this mitochondrial Ca^{2+} uptake in cardiomyocytes possesses a very low MCU current and constitutes less than 1% of total Ca^{2+} uptake [88, 90], it plays a key role in coordinating between excitation and metabolism coupling [91]. In a healthy heart, two models of mitochondrial Ca^{2+} dynamics have been suggested by Cao et al., the first model suggests that Ca^{2+} concentration oscillates in a beat-to-beat manner in cardiomyocytes whereas, the second model emphasizes gradual Ca^{2+} uptake by cardiac mitochondria. On the contrary, in the damaged heart the Ca^{2+} mishandling within the mitochondria is well documented [92]. The MICU1 protein content is significantly low following I/R due to inhibition of translocase expression of the outer membrane. Furthermore, treatment attempts using siRNA on myocardial MICU1 aggravated the ischemic episode increasing tissue damage and depressing cardiac function due to apparent Ca^{2+} overload [86].

As it is quite evident that the uncontrolled influx of Ca^{2+} is disastrous for cardiomyocytes, and MCU is the major route for Ca^{2+} entry. Therefore, alteration in the expression of MCU can be a promising target for cardioprotection. For the inhibition of MCU ruthenium red and its derivatives are generally used, however, ruthenium red has nonspecific activity towards other ion channels [93] as well which does not make it a suitable inhibitor and prevents it from usage as a therapeutic agent. Recently, two new highly selective MCU inhibitors were developed one is DS16570511 prevents Ca^{2+} overload and raises cardiac contractility without affecting heart rate [94]. The second one is Ru265 is negligibly toxic and prevents hypoxia in the cell model [95]. Mitoxanthrone, an anticancer drug that showed its efficiency in inhibiting MCU [96], similarly kaempferol known as an anticancer [97] and cardioprotective drug [98] could prevent Ca^{2+} created arrhythmias [99]. These can be promising drugs in preventing Ca^{2+} related risks to cardiomyocytes but they require more animal study, and careful modeling and validation before adapting as therapeutics.

5.4 Mitochondrial potassium channel

On one side where the opening of mitochondrial mega channels like MPTP and Ca^{2+} uniporter represents a hallmark of cell death, on the other hand, the transport of

K^+ through ion channels is known to play a central role in neural and cardioprotection [100–102]. The membrane potential and permeability of the inner mitochondrial membrane are strictly controlled for efficient ATP production. The presence of an electrophoretic pathway for entry and antiporter mechanism for the exit of K^+ has been well established and they critically regulate the mitochondrial volume and function. The transport of K^+ ions from the cytosol to the mitochondrial matrix is carefully conducted through ion channels by utilizing electrogenic transport, where the proton ejection by the electron transport system generates enough membrane potential for the influx of K^+. There are four kinds of mitochondrial K^+ channels (**Figure 2**) present in the inner membrane the ATP regulated [103], Ca^{2+} regulated [104], Twin pore TASK channel [105], and voltage-gated Kv1.3 potassium channel [106]. These channels resemble the plasma membrane potassium channels in their basic biophysical properties and are regulated to avoid the membrane potential collapse.

5.4.1 ATP sensitive potassium channel

Several mitochondrial K^+ channels (**Figure 2**) have been discovered so far but ATP sensitizing uptake of K^+ has gazed maximum attention. The cardiac ischemic conditioning was first believed to be working on K_{ATP} of the plasma membrane counterpart but based on pharmacological analysis with channel openers and inhibitors shown to affect mitochondrial K_{ATP} channel. The mitoK_{ATP} channel was first identified in rat liver mitochondria using the patch-clamp method [90] and was later found in the inner mitochondrial membrane [107, 108]. They are situated at the crossroad of metabolism and membrane sensitivity. The molecular identities of a mitoK_{ATP} were recently determined by Angela Paggio et al. [109], which is similar to its plasma membrane counterparts that consisting of pore-forming potassium channel CCDC51 (MITOK) and ATP-binding cassette (ABC) transporter ABCB8 (MITOSUR); however, the detailed assembly and function mechanism is still unknown due to the missing of structural information. The plasma membrane K_{ATP} is heterooctameric, containing four inward rectifying potassium channel subunits of Kir6.1 and four

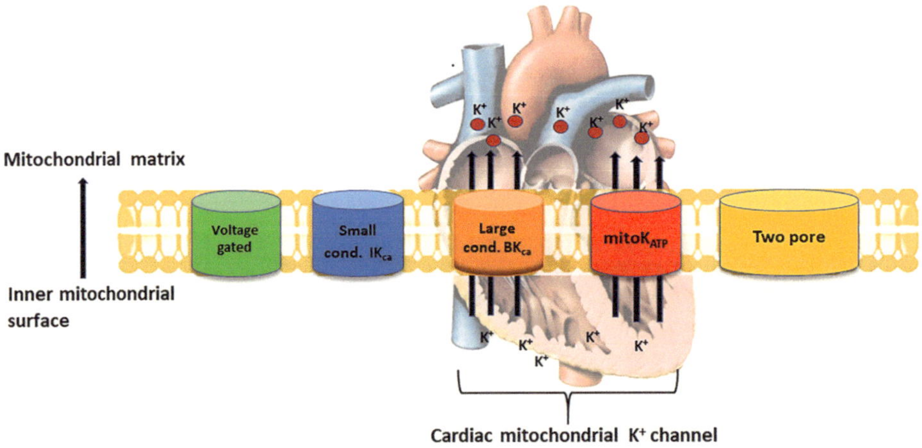

Figure 2.
Mitochondrial potassium channels (a) voltage-gated potassium channels (Kv 1.3, Kv 1.1, Kv 1.5), (b) small and large conductance mitoK_{ATP} channel (IK_{CA}, BK_{Ca}), (c) ATP sensitive K^+ channel, (d) twin pore K^+ channel. mitoK_{ATP} and mitoBK_{Ca} having extensively studied for cardioprotection.

sulfonylurea receptor subunits, which belong to the ABC transporter family [110]. Whether this newly identified mitoK_{ATP} is occupying a similar octameric assembly as the plasma membrane K_{ATP} is still unknown and moreover, it may not present the only version of mitoK_{ATP} as channels such as Kir6.1 has also been suggested in the formation of mitoK_{ATP} [111]. Nevertheless, they are believed to play a central role in cardioprotection, since the bizarre method of cardioprotection called ischemic conditioning was introduced. Ischemic preconditioning was first observed with plasma membrane using K_{ATP} channels as effectors, but later mitochondrial potassium channel became an interesting target for the same. The pathway involves activating the protein kinase and generating ROS, but the precise role of mitoK_{ATP} is not very well established. Therefore, evidence of molecular structure for the mitoK_{ATP} will subside the pharmacology-based arguments of its existence and role in the process of preconditioning. According to a study by Peng Duan et al., mitoK_{ATP} channel opening is helpful in the optimal expression of protein kinase B (p-AKT) and forkhead box protein O1 (pFoxo1) in an insulin-resistant cell. The increased p-Foxo1 is phosphorylated by p-AKT, reducing its transcriptional efficiency, and transferred out of the nucleus, and prevents the expression of pro-apoptotic protein, thereby preventing apoptosis [112].

A study done by Garlid et al. in 2006, showed that the K^+ ion uptake in the mitochondria leads to increased ROS production as they explain that the opening of the mitoK_{ATP} channel will lead to a small amount of K^+ uptake but this lowered potential will increase the matrix volume and pH by a persistent steady state. Valinomycin was used to induce the mitoK_{ATP} opener and an increased pH caused an increased ROS production, when the acetic acid influx increased for compensation of the alkaline matrix the ROS production also reduced proving that the alkalinity is a cause for ROS production. It was also proved by them that the ROS was generated from the complex I of the electron transport system [113].

5.4.2 Ca^{2+} activated potassium channel

These channels were first discovered in the glioma cell line LN-229 and have been extensively studied in the brain and cardiac cells since then [104]. The calcium-activated potassium channel is of two types small and intermediate conductance K^+ channel and large or big conductance K^+ channel. The small and intermediate channels are only calcium-dependent and not voltage-dependent they possess calmodulin for the Ca^{2+} binding at the C-terminal region. Their major function is promoting proliferation and migration of dendritic cells and smooth muscle [114].

The first evidence of its presence in the inner mitochondrial membrane was found in the late 90s by Siemen and coworkers although showing that it possesses a conductance of 300 pS [104]. Its role in protecting the heart from ischemic insult was first discovered by Xu et al. and their structural characterization in the plasma membrane indicates that it originates from potassium calcium-activated channel subfamily M alpha 1 (Kcnma1) gene containing extracellular N terminus and intracellular C-terminus [115]. It has been proven in 2013 that the BK_{Ca} pore-forming alpha subunits are encoded by the same genes (Kcnma1) as the basis of why they possess the same physiological properties [116].

The big conductance Ca^{2+} sensitive K^+ channel also known as MitoBK_{Ca} on the other hand is intuitive to voltage and mechanical stress alongside Ca^{2+} sensitivity. The knockout experiments have proven for the MitoBK_{Ca} channels to have a cardioprotective effect by reversing the ROS production and opening MPTP. It has also been shown that these channels form a multiprotein complex with several proteins involved in apoptotic machinery [117].

mitoBK_{Ca} channel [118] represent themselves as a key pathophysiological target due to their sensitivity towards calcium, voltage, and a range of cellular components. Several small molecular openers for BK_{Ca} and pharmacological agents have provided very insightful information to decipher the role of BK_{Ca} channel. Pharmacological agents like NS1619 and NS11021 have been used to activate BK_{Ca} can potentially play a vital role in cardioprotection. However, they fail to reach the clinical applications due to their non-specificity [119–122]. Although it is posing a great deal of difficulty in developing the BK_{Ca} activators, it becomes essential considering that expression of BK_{Ca} is vital for cardioprotection [116, 123].

The first representation BK_{Ca} playing an essential role in cardioprotection from I/R injury was performed by using NS1619, whose effect was blocked by praxilline [115]. A 3 mM NS1619 preconditioning showed an improved reduction of infarct size, possibly by modulated Ca^{2+} and ROS concentration [117]. BK_{Ca} mediated cardioprotection involves ROS, Ca^{2+}, and MPTP and their interplay. It is anticipated that reduction of deleterious ROS through BK_{Ca} activation prevents the excess release of Ca^{2+} from the endoplasmic reticulum subsequently reducing the influx and overload in mitochondria preventing the cell from injury.

5.4.3 Voltage-gated potassium channel

These are the most diverse family of K^+ channels. They are grouped into 12 families comprising 40 of the 90 genes present in human cells [124]. These channels mostly consist of six transmembrane helices (S1-S6) where two of them (S5-S6) form the loop and four of them are proceedings to the loop (S1-S4). The fourth positively charged loop senses the change in the membrane potential. These channels have a wide variety and therefore, they represent a fine regulation of K^+ flux in the homeostasis and pathological processes. The mitochondrial counterpart mitoKv1.3 is found in lymphocytes [125] and many carcinogenic cells [126, 127], they present similar physiological functions as the plasma membrane counterpart and, they are translated from the same gene. mitoKv1.3 is a target for pro-apoptotic protein Bax [128]. Their complex prevents the opening of mitoKv1.3 channel for K^+ influx and therefore causes the disturbance in membrane potential and eventually leads to apoptotic cell death. Therefore, mitoKv1.3 has represented itself as a new tool for targeted cell death for many tumor cells by triggering mitochondria-induced apoptosis. Their presence in cardiac cells has not been reported. Similar to Kv1.3 other potassium channels like Kv1.1 and Kv1.5 have also shown dual origin in both mitochondrial and plasma membrane causing cell apoptosis by targeting macrophages [129, 130].

5.4.4 Twin pore potassium channel

The mitochondrial TASK-3 was discovered in human keratinocyte HeCaT cells using the patch-clamp method [131]. It shows similarity with its plasma membrane counterpart and its activity is inhibited at acidic pH [132]. Lidocaine and low pH completely block the task channel activity in mitochondria. TASK-3 is essential for the survival of WM35 melanoma cells [133] but its activity in the mitochondrial dysfunction in cardiac reperfusion is not known.

6. Mitochondria channels as a therapeutic target of heart failure

The above discussions have made it clear that since the inner mitochondrial channels regulate the onset of apoptosis and cell death, they present an important target for cardioprotective therapeutics. Not only mitochondrial potassium channel but also MPTP, MCU, connexin-43, and protein uncoupling have shown their potential roles in reducing myocardial infarct size and preventing heart failure.

A sudden opening of MPTP can be triggered through a high concentration of Ca^{2+}, high amount of ROS production, and decrease in mitochondrial membrane potential, which results in the loss of proton gradient appearing as an uncoupling effect, which prevents ATP formation and promotes its hydrolysis [40]. Subsequently, the proton gradient utilizes the Ca^{2+} uptake and causes the matrix swelling as an approach of the MPTP to prevent the detrimental rise in Ca^{2+} in the mitochondrial matrix [134]. It is known that opening of MPTP for a short duration can proceed without affecting cell viability and can also contribute to cardioprotection through participating in pre ischemic conditioning and it later prevents the opening of MPTP pore during ischemic reperfusion preventing cell damage and the onset of heart failure [17]. Although the evidence to support its cardioprotective functions is majorly based on the pharmacology and genetic observation that avoided MPTP opening. It has been shown that the administration of cyclosporin A shows a cardioprotective effect by preventing MPTP opening in the mice model; however, the results are mixed for the large mammalian model [135].

As mentioned earlier an increase in the Ca^{2+} concentration contributes to the opening of the MPTP channel, it is also necessary to mention that the Ca^{2+} is essential for the key enzyme activation in the oxidation of the substrates that fuel the respiratory chain, followed by ATP formation. It is very unfortunate that despite the advancement in technologies we are still unable to determine the physiological and pathological concentration of Ca^{2+} in the mitochondrial matrix [136]. Nevertheless, the Ca^{2+} homeostasis is maintained within the mitochondrial matrix by the uptake of Ca^{2+} through uniporters and the release is catalyzed by Na^+/Ca^{2+} exchanger. The understanding of the molecular nature of Ca^{2+} uniporter has advanced our knowledge about Ca^{2+} homeostasis. The deletion of mitochondrial calcium uniporter gene from the embryonic and adult mice has shown completely contradicting results and the reasons of which have not yet been fully understood. However, the results that appeared in adult mice fully support the role of calcium overload leading to MPTP opening and eventually cell death. This is further supported by the $Na^+/Ca^{2+}/Li^+$ exchanger knockout mice which showed the overload of Ca^{2+} leading to MPTP opening on the onset of ischemic reperfusion leading to cell death. Therefore, the drugs that intend to target Ca^{2+} uniporter for therapeutics need to validate the contrast effects before large animal and clinical testing [137].

Connexin43(Cx43) is a well-known channel for the intercellular connections by forming the gap junctions, but apart from the plasma membrane occurrence, they are also known to be present in cellular organelle like subsarcolemmal mitochondria [138], nucleus [139], and exosomes [140]. Cx43 plays an important role in ischemic-reperfusion injury and its prevention. According to pharmacological evidence the concentration of Cx43 increases with the introduction of diazoxide DZX or fibroblast growth factor 2 to prevent myocardial injury, but this increase is not observed after the ischemic preconditioning protocol [141]. It also interacts with the mitochondrial potassium channel [141] and regulates nitric oxide formation. The role of Cx43 is certain in cardioprotection but the exact mechanism and function remain to be elucidated.

At last, the presence of several mitochondrial K^+ channels and their activity in the failing heart presents them as a crucial target in the therapeutics of myocardial dysfunction. The K^+ uptake and release play a central role in the maintenance of mitochondrial matrix volume. The electrophoretic influx of K^+ is balanced by the K^+/H^+ antiporter [142]. Valinomycin triggers the uncontrolled K^+ influx disturbing the mitochondrial polarization and causing the swelling of the matrix. The function of potassium channels is basically to maintain the matrix volume. Initially, surface K^+ channel was suggested to play an essential role in ischemic pre and postconditioning but later when diazoxide (DXZ), which was involved in cardioprotection of non-contractable heart, did not show any effect on surface K^+ channel, whereas drugs that were only targeting surface K^+ did not show cardioprotective effect. On the contrary, the isolated mitochondria showed restored activity of ATP inhibited flux and showed inhibition caused by 5-hydroxydecanoate (5HD) [143]. This shifts the attention to the mitochondrial K channel for cardioprotection but ever since the DXZ and 5HD also affect mitochondrial physiology in general it requires the molecular structure information and in vivo attempts for concrete statements. The structural information will provide tools for determining its exact function in myocardial ischemia/reperfusion in a failing heart, Ca^{2+} transport and MPTP opening, and protein involved in ROS formation, followed by improving therapeutic approaches [144]. Similar to the ATP activated K^+ channel, Ca^{2+} and voltage-activated channels are also pharmacologically proven to play a similar role as mitoK_{ATP}, the ischemic conditioning protocol triggers the formation of protein kinase C which is helpful in an increased opening mitoK_{Ca} channel.

7. Current progress in the field

The strategies used to protect the heart from opening MPTP and mitochondrial calcium uniporter pores and in the case of ischemia conditioning, the opening of mitoK_{ATP} and BK_{Ca} channel plays a vital role in the cardioprotection. CsA is a well-known desensitizer for MPTP, but it did not prove to be the best option in clinical trials [65]. CsA exerts its activity by binding to CyPD, but in cases of intense stimuli, the pore opening becomes independent of CyPD. Therefore, there is a need of developing more pharmacological agents that can directly inhibit the MPTP openings, but they require further information about the structural insight of the pore. Although CyPD activates pore opening the complete mechanism is still unclear. Similarly, although it is evident that ROS and Ca^{2+} influence the MPTP opening but without knowing the structural details of MPTP we fail to conclude how they do so.

Mitochondrial calcium regulates a range of myocyte functions alongside energy production like cell division and trophism. With the development of MCU structures over the years, they have emerged as a very important target for cardioprotection, but the development of a reliable drug is still in process. Potassium channels are also widely accepted as an important target and are closely linked to modulating the apoptotic process. This information is present due to the pharmacology of the channel openers and inhibitors. Very limited knowledge is present to show concrete evidence. The molecular structure can be helpful in understanding and curing several mitochondria-associated diseases.

The inhibition of MPTP channel opening and the mitoK channel both elicit the cardioprotection and are likely to be related. Uptake of K^+ through mitoK decreased the mitochondrial membrane potential which reduces the mitochondrial Ca^{2+}, which in turn decreases the possibility of MPTP opening. Along with ATP production, and ROS regulation, mitochondrial channels like MPTP, Ca^{2+} channel, and mitoK channels are established to play a crucial role in cardioprotection. The mechanism, however, for their connection and coordination with each other in the process of cardioprotection is far from conclusive.

Recent attempts to translate cardioprotective strategies that target some of these mitochondrial ion channels have been hugely disappointing, and the translational of these strategies in clinical settings have not been successful. Several drugs have been tested on various animal models that have shown certain cardioprotective mechanisms. However, the lack of knowledge about the underlying mechanism of protective actions needs a lot of following studies to design modulators specific for mitochondrial channels with regards to cardioprotection in human trials. In the light of studies available, we still have a long way to go in the depth of the cardioprotective mechanism.

8. Conclusions

Conclusively, heart failure is an outcome of cardiac injury that originated due to a variety of etiologies and denotes a complex clinical syndrome. Several mitochondrial channels associated mechanisms have been recognized that drive the depletion of cardiomyocytes before cell death. These observations not only provide a link of overall heart health with mitochondrial channel opening and closing but also inspires therapeutic approaches. The core molecular identity of some mitochondrial channels like MCU and mitoK_{ATP} are discovered recently, whereas most mitochondrial potassium channels are in their intermediate state. These channels act as switches to control the development of ischemic injury either towards recovery or the loss of viability. The progress towards understanding the molecular identity and mechanism of channel opening and inhibition will help to translate the experimental approaches into promising therapeutic development to combat a deadly health concern.

Acknowledgements

We thank Dr. Nileshkumar Dubey for his help with the figures. **Figure 1** was created with Biorender.com.

Conflict of interest

The authors declare no conflict of interest.

Author details

Keerti Mishra and Min Luo*
Department of Biological Sciences, National University of Singapore, Singapore

*Address all correspondence to: dbslmin@nus.edu.sg

References

[1] Savarese G, Lund LH. Global public health burden of heart failure. Cardiac Failure Review. 2017;**3**(1):7-11. DOI: 10.15420/cfr.2016:25:2

[2] Marin-Garcia J, Goldenthal MJ, Moe GW. Mitochondrial pathology in cardiac failure. Cardiovascular Research. 2001;**49**(1):17-26. DOI: 10.1016/s0008-6363(00)00241-8

[3] Russell LK, Finck BN, Kelly DP. Mouse models of mitochondrial dysfunction and heart failure. Journal of Molecular and Cellular Cardiology. 2005;**38**(1): 81-91. DOI: 10.1016/j.yjmcc.2004.10.010

[4] Zorov DB, Juhaszova M, Sollott SJ. Mitochondrial reactive oxygen species (ROS) and ROS-induced ROS release. Physiological Reviews. 2014;**94**(3):909-950. DOI: 10.1152/physrev.00026.2013

[5] Checchetto V, Azzolini M, Peruzzo R, Capitanio P, Leanza L. Mitochondrial potassium channels in cell death. Biochemical and Biophysical Research Communications. 2018;**500**(1):51-58. DOI: 10.1016/j.bbrc.2017.06.095

[6] Törnroth-Horsefield S, Neutze R. Opening and closing the metabolite gate. Proceedings of the National Academy of Sciences of the United States of America. 2008;**105**(50):19565-19566

[7] Aubert G, Vega RB, Kelly DP. Perturbations in the gene regulatory pathways controlling mitochondrial energy production in the failing heart. Biochimica et Biophysica Acta. 2013;**1833**(4):840-847. DOI: 10.1016/j.bbamcr.2012.08.015

[8] Margulis L, Sagan D. Origins of Sex: Three Billion Years of Genetic Recombination. New Haven, Connecticut: Yale University Press; 1990

[9] Fozzard HA, Haber E, Jennings RB, Katz AM. The heart and cardiovascular system. New York: Raven Press; 1986

[10] Escobales N, Nunez RE, Jang S, Parodi-Rullan R, Ayala-Pena S, Sacher JR, et al. Mitochondria-targeted ROS scavenger improves post-ischemic recovery of cardiac function and attenuates mitochondrial abnormalities in aged rats. Journal of Molecular and Cellular Cardiology. 2014;**77**:136-146. DOI: 10.1016/j.yjmcc.2014.10.009

[11] Glancy B, Willis WT, Chess DJ, Balaban RS. Effect of calcium on the oxidative phosphorylation cascade in skeletal muscle mitochondria. Biochemistry. 2013;**52**(16):2793-2809

[12] Boyman L, Williams GS, Khananshvili D, Sekler I, Lederer W. NCLX: The mitochondrial sodium calcium exchanger. Journal of Molecular and Cellular Cardiology. 2013;**59**:205-213

[13] Birk AV, Chao WM, Bracken C, Warren JD, Szeto HH. Targeting mitochondrial cardiolipin and the cytochrome c/cardiolipin complex to promote electron transport and optimize mitochondrial ATP synthesis. British Journal of Pharmacology. 2014;**171**(8):2017-2028. DOI: 10.1111/bph.12468

[14] Jaswal JS, Keung W, Wang W, Ussher JR, Lopaschuk GD. Targeting fatty acid and carbohydrate oxidation—a novel therapeutic intervention in the ischemic and failing heart. Biochimica et Biophysica Acta (BBA) - Molecular Cell Research. 2011;**1813**(7):1333-1350

[15] Shao D, Villet O, Zhang Z, Choi SW, Yan J, Ritterhoff J, et al. Glucose promotes cell growth by

suppressing branched-chain amino acid degradation. Nature Communications. 2018;**9**(1):2935. DOI: 10.1038/s41467-018-05362-7

[16] Bauer TM, Murphy E. Role of mitochondrial calcium and the permeability transition pore in regulating cell death. Circulation Research. 2020;**126**(2):280-293. DOI: 10.1161/CIRCRESAHA.119.316306

[17] Hausenloy D, Wynne A, Duchen M, Yellon D. Transient mitochondrial permeability transition pore opening mediates preconditioning-induced protection. Circulation. 2004;**109**(14):1714-1717. DOI: 10.1161/01.CIR.0000126294.81407.7D

[18] Costantini P, Chernyak BV, Petronilli V, Bernardi P. Selective inhibition of the mitochondrial permeability transition pore at the oxidation-reduction sensitive dithiol by monobromobimane. FEBS Letters. 1995;**362**(2):239-242

[19] Bernardi P. Mitochondrial transport of cations: Channels, exchangers, and permeability transition. Physiological Reviews. 1999;**79**(4):1127-1155. DOI: 10.1152/physrev.1999.79.4.1127

[20] Halestrap AP, Clarke SJ, Javadov SA. Mitochondrial permeability transition pore opening during myocardial reperfusion—A target for cardioprotection. Cardiovascular Research. 2004;**61**(3):372-385

[21] Murry CE, Jennings RB, Reimer KA. Preconditioning with ischemia: A delay of lethal cell injury in ischemic myocardium. Circulation. 1986;**74**(5):1124-1136. DOI: 10.1161/01.cir.74.5.1124

[22] Edwards RJ, Saurin AT, Rakhit RD, Marber MS. Therapeutic potential of ischaemic preconditioning. British Journal of Clinical Pharmacology. 2000;**50**(2):87-97. DOI: 10.1046/j.1365-2125.2000.00236.x

[23] Yellon DM, Downey JM. Preconditioning the myocardium: From cellular physiology to clinical cardiology. Physiological Reviews. 2003;**83**(4):1113-1151. DOI: 10.1152/physrev.00009.2003

[24] Heusch G. Molecular basis of cardioprotection. Circulation Research. 2015;**116**(4):674-699. DOI: 10.1161/CIRCRESAHA.116.305348

[25] Murriel CL, Mochly-Rosen D. Opposing roles of δ and εPKC in cardiac ischemia and reperfusion: Targeting the apoptotic machinery. Archives of Biochemistry and Biophysics. 2003;**420**(2):246-254

[26] Heusch G, Boengler K, Schulz R. Cardioprotection: Nitric oxide, protein kinases, and mitochondria. Circulation 2008;**118**(19):1915-1919. DOI: 10.1161/CIRCULATIONAHA.108.805242

[27] Baines CP, Song C-X, Zheng Y-T, Wang G-W, Zhang J, Wang O-L, et al. Protein kinase Cε interacts with and inhibits the permeability transition pore in cardiac mitochondria. Circulation Research. 2003;**92**(8):873-880

[28] Ylitalo K, Ala-Rämi A, Vuorinen K, Peuhkurinen K, Lepojärvi M, Kaukoranta P, et al. Reversible ischemic inhibition of F1F0-ATPase in rat and human myocardium. Biochimica et Biophysica Acta - Bioenergetics. 2001;**1504**(2-3):329-339

[29] Chen Z, Chua CC, Ho Y-S, Hamdy RC, Chua BH. Overexpression of Bcl-2 attenuates apoptosis and protects against myocardial I/R injury in transgenic mice. American Journal of Physiology-Heart and Circulatory Physiology. 2001;**280**(5):H2313-H2H20

[30] Hattori R, Hernandez TE, Zhu L, Maulik N, Otani H, Kaneda Y, et al. An essential role of the antioxidant gene Bcl-2 in myocardial adaptation to ischemia: An insight with antisense Bcl-2 therapy. Antioxidants & Redox Signaling. 2001;**3**(3):403-413. DOI: 10.1089/15230860152409059

[31] Di Lisa F, Bernardi P. Mitochondria and ischemia–reperfusion injury of the heart: Fixing a hole. Cardiovascular Research. 2006;**70**(2):191-199

[32] Weiss JN, Korge P, Honda HM, Ping P. Role of the mitochondrial permeability transition in myocardial disease. Circulation Research. 2003;**93**(4):292-301. DOI: 10.1161/01.RES.0000087542.26971.D4

[33] Hausenloy DJ, Maddock HL, Baxter GF, Yellon DM. Inhibiting mitochondrial permeability transition pore opening: A new paradigm for myocardial preconditioning? Cardiovascular Research. 2002;**55**(3):534-543. DOI: 10.1016/s0008-6363(02)00455-8

[34] Javadov SA, Clarke S, Das M, Griffiths EJ, Lim KH, Halestrap AP. Ischaemic preconditioning inhibits opening of mitochondrial permeability transition pores in the reperfused rat heart. The Journal of Physiology. 2003;**549**(Pt 2):513-524. DOI: 10.1113/jphysiol.2003.034231

[35] Garlid KD, Paucek P, Yarov-Yarovoy V, Murray HN, Darbenzio RB, D'Alonzo AJ, et al. Cardioprotective effect of diazoxide and its interaction with mitochondrial ATP-sensitive K+ channels. Possible mechanism of cardioprotection. Circulation Research. 1997;**81**(6):1072-1082. DOI: 10.1161/01.res.81.6.1072

[36] Oldenburg O, Cohen MV, Yellon DM, Downey JM. Mitochondrial KATP channels: Role in cardioprotection. Cardiovascular Research. 2002;**55**(3):429-437

[37] O'Rourke B. Myocardial KATP channels in preconditioning. Circulation Research. 2000;**87**(10):845-855

[38] Kowaltowski AJ, Seetharaman S, Paucek P, Garlid KD. Bioenergetic consequences of opening the ATP-sensitive K(+) channel of heart mitochondria. American Journal of Physiology. Heart and Circulatory Physiology. 2001;**280**(2):H649-H657. DOI: 10.1152/ajpheart.2001.280.2.H649

[39] Haworth RA, Hunter DR. The Ca2+−induced membrane transition in mitochondria: II. Nature of the Ca2+ trigger site. Archives of Biochemistry and Biophysics. 1979;**195**(2):460-467

[40] Bernardi P, Di Lisa F. The mitochondrial permeability transition pore: molecular nature and role as a target in cardioprotection. Journal of Molecular and Cellular Cardiology. 2015;**78**:100-106

[41] Bernardi P, Krauskopf A, Basso E, Petronilli V, Blachly-Dyson E, Di Lisa F, et al. The mitochondrial permeability transition from in vitro artifact to disease target. The FEBS Journal. 2006;**273**(10):2077-2099. DOI: 10.1111/j.1742-4658.2006.05213.x

[42] Rasola A, Bernardi P. Mitochondrial permeability transition in Ca2+−dependent apoptosis and necrosis. Cell Calcium. 2011;**50**(3):222-233

[43] Petronilli V, Penzo D, Scorrano L, Bernardi P, Di Lisa F. The mitochondrial permeability transition, release of cytochrome c and cell death. Correlation with the duration of pore openings in situ. The Journal of Biological Chemistry. 2001;**276**(15):12030-12034. DOI: 10.1074/jbc.M010604200

[44] HüSER J, Blatter LA. Fluctuations in mitochondrial membrane potential caused by repetitive gating of the permeability transition pore. Biochemical Journal. 1999;**343**(2):311-317

[45] Crompton M, Ellinger H, Costi A. Inhibition by cyclosporin A of a Ca2+–dependent pore in heart mitochondria activated by inorganic phosphate and oxidative stress. The Biochemical Journal. 1988;**255**(1):357-360

[46] Hüser J, Rechenmacher CE, Blatter LA. Imaging the permeability pore transition in single mitochondria. Biophysical Journal. 1998;**74**(4): 2129-2137

[47] Halestrap AP. What is the mitochondrial permeability transition pore? Journal of Molecular and Cellular Cardiology. 2009;**46**(6):821-831

[48] Hunter DR, Haworth RA. The Ca2+–induced membrane transition in mitochondria. I. The protective mechanisms. Archives of Biochemistry and Biophysics. 1979;**195**(2):453-459. DOI: 10.1016/0003-9861(79)90371-0

[49] Halestrap AP, Brenner C. The adenine nucleotide translocase: A central component of the mitochondrial permeability transition pore and key player in cell death. Current Medicinal Chemistry. 2003;**10**(16):1507-1525. DOI: 10.2174/0929867033457278

[50] McStay GP, Clarke SJ, Halestrap AP. Role of critical thiol groups on the matrix surface of the adenine nucleotide translocase in the mechanism of the mitochondrial permeability transition pore. The Biochemical Journal. 2002;**367**(Pt 2):541-548. DOI: 10.1042/BJ20011672

[51] Woodfield K, Rück A, Brdiczka D, Halestrap AP. Direct demonstration of a specific interaction between cyclophilin-D and the adenine nucleotide translocase confirms their role in the mitochondrial permeability transition. Biochemical Journal. 1998;**336**(2):287-290

[52] Pebay-Peyroula E, Dahout-Gonzalez C, Kahn R, Trezeguet V, Lauquin GJ, Brandolin G. Structure of mitochondrial ADP/ATP carrier in complex with carboxyatractyloside. Nature. 2003;**426**(6962):39-44. DOI: 10.1038/nature02056

[53] Leung AW, Varanyuwatana P, Halestrap AP. The mitochondrial phosphate carrier interacts with cyclophilin D and may play a key role in the permeability transition. Journal of Biological Chemistry. 2008;**283**(39):26312-26323

[54] Brenner C, Cadiou H, Vieira HL, Zamzami N, Marzo I, Xie Z, et al. Bcl-2 and Bax regulate the channel activity of the mitochondrial adenine nucleotide translocator. Oncogene. 2000;**19**(3):329-336. DOI: 10.1038/sj.onc.1203298

[55] Marzo I, Brenner C, Zamzami N, Susin SA, Beutner G, Brdiczka D, et al. The permeability transition pore complex: A target for apoptosis regulation by caspases and Bcl-2–related proteins. Journal of Experimental Medicine. 1998;**187**(8):1261-1271

[56] Neginskaya MA, Solesio ME, Berezhnaya EV, Amodeo GF, Mnatsakanyan N, Jonas EA, et al. ATP synthase C-subunit-deficient mitochondria have a small cyclosporine A-sensitive channel, but lack the permeability transition pore. Cell Reports. 2019;**26**(1):11-7.e2

[57] Beutner G, Alanzalon RE, Porter GA. Cyclophilin D regulates the dynamic assembly of mitochondrial ATP synthase

into synthasomes. Scientific Reports. 2017;**7**(1):1-12

[58] Gateau-Roesch O, Argaud L, Ovize M. Mitochondrial permeability transition pore and postconditioning. Cardiovascular Research. 2006;**70**(2):264-273. DOI: 10.1016/j.cardiores.2006.02.024

[59] Khaliulin I, Clarke SJ, Lin H, Parker J, Suleiman MS, Halestrap AP. Temperature preconditioning of isolated rat hearts–a potent cardioprotective mechanism involving a reduction in oxidative stress and inhibition of the mitochondrial permeability transition pore. The Journal of Physiology. 2007;**581**(3):1147-1161

[60] Javadov S, Choi A, Rajapurohitam V, Zeidan A, Basnakian AG, Karmazyn M. NHE-1 inhibition-induced cardioprotection against ischaemia/reperfusion is associated with attenuation of the mitochondrial permeability transition. Cardiovascular Research. 2008;**77**(2):416-424

[61] Adlam VJ, Harrison JC, Porteous CM, James AM, Smith RA, Murphy MP, et al. Targeting an antioxidant to mitochondria decreases cardiac ischemia-reperfusion injury. The FASEB Journal. 2005;**19**(9):1088-1095. DOI: 10.1096/fj.05-3718com

[62] Javadov SA, Lim KH, Kerr PM, Suleiman M-S, Angelini GD, Halestrap AP. Protection of hearts from reperfusion injury by propofol is associated with inhibition of the mitochondrial permeability transition. Cardiovascular Research. 2000;**45**(2):360-369

[63] Townsend PA, Davidson SM, Clarke SJ, Khaliulin I, Carroll CJ, Scarabelli TM, et al. Urocortin prevents mitochondrial permeability transition in response to reperfusion injury indirectly by reducing oxidative stress. American Journal of Physiology. Heart and Circulatory Physiology. 2007;**293**(2):H928-H938. DOI: 10.1152/ajpheart.01135.2006

[64] Kerr PM, Suleiman MS, Halestrap AP. Reversal of permeability transition during recovery of hearts from ischemia and its enhancement by pyruvate. The American Journal of Physiology. 1999;**276**(2):H496-H502. DOI: 10.1152/ajpheart.1999.276.2.H496

[65] Piot C, Croisille P, Staat P, Thibault H, Rioufol G, Mewton N, et al. Effect of cyclosporine on reperfusion injury in acute myocardial infarction. The New England Journal of Medicine. 2008;**359**(5):473-481. DOI: 10.1056/NEJMoa071142

[66] Basso E, Fante L, Fowlkes J, Petronilli V, Forte MA, Bernardi P. Properties of the permeability transition pore in mitochondria devoid of Cyclophilin D. The Journal of Biological Chemistry. 2005;**280**(19):18558-18561. DOI: 10.1074/jbc.C500089200

[67] Sorgato MC, Keller BU, Stühmer W. Patch-clamping of the inner mitochondrial membrane reveals a voltage-dependent ion channel. Nature. 1987;**330**(6147):498-500

[68] Beavis AD. Properties of the inner membrane anion channel in intact mitochondria. Journal of Bioenergetics and Biomembranes. 1992;**24**(1):77-90

[69] O'Rourke B. Pathophysiological and protective roles of mitochondrial ion channels. The Journal of Physiology. 2000;**529**(1):23-36

[70] Aon MA, Cortassa S, Marbán E, O'Rourke B. Synchronized whole cell oscillations in mitochondrial metabolism

triggered by a local release of reactive oxygen species in cardiac myocytes. Journal of Biological Chemistry. 2003;**278**(45):44735-44744

[71] Aon MA, Cortassa S, Maack C, O'Rourke B. Sequential opening of mitochondrial ion channels as a function of glutathione redox thiol status. Journal of Biological Chemistry. 2007;**282**(30):21889-21900

[72] Zorov DB, Juhaszova M, Sollott SJ. Mitochondrial ROS-induced ROS release: An update and review. Biochimica et Biophysica Acta - Bioenergetics. 2006;**1757**(5-6):509-517

[73] Chance B. The energy-linked reaction of calcium with mitochondria. The Journal of Biological Chemistry. 1965;**240**(6):2729-2748

[74] Gunter TE, Gunter KK, Sheu SS, Gavin CE. Mitochondrial calcium transport: Physiological and pathological relevance. The American Journal of Physiology. 1994;**267**(2 Pt 1):C313-C339. DOI: 10.1152/ajpcell.1994.267.2.C313

[75] Pagliarini DJ, Calvo SE, Chang B, Sheth SA, Vafai SB, Ong SE, et al. A mitochondrial protein compendium elucidates complex I disease biology. Cell. 2008;**134**(1):112-123. DOI: 10.1016/j.cell.2008.06.016

[76] Perocchi F, Gohil VM, Girgis HS, Bao XR, McCombs JE, Palmer AE, et al. MICU1 encodes a mitochondrial EF hand protein required for Ca(2+) uptake. Nature. 2010;**467**(7313):291-296. DOI: 10.1038/nature09358

[77] De Stefani D, Raffaello A, Teardo E, Szabò I, Rizzuto R. A forty-kilodalton protein of the inner membrane is the mitochondrial calcium uniporter. Nature. 2011;**476**(7360):336-340

[78] Baughman JM, Perocchi F, Girgis HS, Plovanich M, Belcher-Timme CA, Sancak Y, et al. Integrative genomics identifies MCU as an essential component of the mitochondrial calcium uniporter. Nature. 2011;**476**(7360):341-345

[79] Pan X, Liu J, Nguyen T, Liu C, Sun J, Teng Y, et al. The physiological role of mitochondrial calcium revealed by mice lacking the mitochondrial calcium uniporter. Nature Cell Biology. 2013;**15**(12):1464-1472. DOI: 10.1038/ncb2868

[80] Rasmussen TP, Wu Y, Joiner ML, Koval OM, Wilson NR, Luczak ED, et al. Inhibition of MCU forces extramitochondrial adaptations governing physiological and pathological stress responses in heart. Proceedings of the National Academy of Sciences of the United States of America. 2015;**112**(29):9129-9134. DOI: 10.1073/pnas.1504705112

[81] Luongo TS, Lambert JP, Yuan A, Zhang X, Gross P, Song J, et al. The mitochondrial calcium uniporter matches energetic supply with cardiac workload during stress and modulates permeability transition. Cell Reports. 2015;**12**(1):23-34. DOI: 10.1016/j.celrep.2015.06.017

[82] Sancak Y, Markhard AL, Kitami T, Kovács-Bogdán E, Kamer KJ, Udeshi ND, et al. EMRE is an essential component of the mitochondrial calcium uniporter complex. Science. 2013;**342**(6164):1379-1382

[83] De Stefani D, Patron M, Rizzuto R. Structure and function of the mitochondrial calcium uniporter complex. Biochimica et Biophysica Acta. 2015;**1853**(9):2006-2011. DOI: 10.1016/j.bbamcr.2015.04.008

[84] Logan CV, Szabadkai G, Sharpe JA, Parry DA, Torelli S, Childs AM, et al.

Loss-of-function mutations in MICU1 cause a brain and muscle disorder linked to primary alterations in mitochondrial calcium signaling. Nature Genetics. 2014;**46**(2):188-193. DOI: 10.1038/ng.2851

[85] Lewis-Smith D, Kamer KJ, Griffin H, Childs AM, Pysden K, Titov D, et al. Homozygous deletion in MICU1 presenting with fatigue and lethargy in childhood. Neurology Genetics. 2016;**2**(2):e59. DOI: 10.1212/NXG.0000000000000059

[86] Luongo TS, Lambert JP, Gross P, Nwokedi M, Lombardi AA, Shanmughapriya S, et al. The mitochondrial Na+/Ca 2+ exchanger is essential for Ca 2+ homeostasis and viability. Nature. 2017;**545**(7652):93-97

[87] Bick AG, Wakimoto H, Kamer KJ, Sancak Y, Goldberger O, Axelsson A, et al. Cardiovascular homeostasis dependence on MICU2, a regulatory subunit of the mitochondrial calcium uniporter. Proceedings of the National Academy of Sciences of the United States of America. 2017;**114**(43):E9096-EE104. DOI: 10.1073/pnas.1711303114

[88] Bers DM. Cardiac excitation–contraction coupling. Nature. 2002;**415**(6868):198-205

[89] Giorgi C, Marchi S, Pinton P. The machineries, regulation and cellular functions of mitochondrial calcium. Nature Reviews Molecular Cell Biology. 2018;**19**(11):713-730

[90] Fieni F, Lee SB, Jan YN, Kirichok Y. Activity of the mitochondrial calcium uniporter varies greatly between tissues. Nature Communications. 2012;**3**(1):1-12

[91] Kwong JQ. The mitochondrial calcium uniporter in the heart: Energetics and beyond. The Journal of Physiology. 2017;**595**(12):3743-3751. DOI: 10.1113/JP273059

[92] Boyman L, Williams GS, Lederer WJ. Mitochondrial calcium and ischemia: Reperfusion injury in heart. In: Hockenbery D. (eds) Mitochondria and Cell Death. Cell Death in Biology and Disease. New York: Humana press; 2016. pp. 17-43

[93] Kamer KJ, Mootha VK. The molecular era of the mitochondrial calcium uniporter. Nature Reviews. Molecular Cell Biology. 2015;**16**(9):545-553. DOI: 10.1038/nrm4039

[94] Kon N, Murakoshi M, Isobe A, Kagechika K, Miyoshi N, Nagayama T. DS16570511 is a small-molecule inhibitor of the mitochondrial calcium uniporter. Cell Death Discovery. 2017;**3**(1):1-7

[95] Woods JJ, Nemani N, Shanmughapriya S, Kumar A, Zhang M, Nathan SR, et al. A selective and cell-permeable mitochondrial calcium uniporter (MCU) inhibitor preserves mitochondrial bioenergetics after hypoxia/reoxygenation injury. ACS Central Science. 2019;**5**(1):153-166. DOI: 10.1021/acscentsci.8b00773

[96] Arduino DM, Wettmarshausen J, Vais H, Navas-Navarro P, Cheng Y, Leimpek A, et al. Systematic identification of MCU modulators by orthogonal interspecies chemical screening. Molecular Cell. 2017;**67**(4):711-23 e7. DOI: 10.1016/j.molcel.2017.07.019

[97] Imran M, Rauf A, Shah ZA, Saeed F, Imran A, Arshad MU, et al. Chemo-preventive and therapeutic effect of the dietary flavonoid kaempferol: A comprehensive review. Phytotherapy Research. 2019;**33**(2):263-275. DOI: 10.1002/ptr.6227

[98] Imran M, Salehi B, Sharifi-Rad J, Aslam Gondal T, Saeed F, Imran A,

et al. Kaempferol: A key emphasis to its anticancer potential. Molecules. 2019;**24**(12):2277. DOI: 10.3390/molecules24122277

[99] Schweitzer MK, Wilting F, Sedej S, Dreizehnter L, Dupper NJ, Tian Q, et al. Suppression of arrhythmia by enhancing mitochondrial Ca(2+) uptake in catecholaminergic ventricular tachycardia models. JACC: Basic to Translational Science. 2017;**2**(6):737-747. DOI: 10.1016/j.jacbts.2017.06.008

[100] Szewczyk A, Marban E. Mitochondria: A new target for K channel openers? Trends in Pharmacological Sciences. 1999;**20**(4):157-161. DOI: 10.1016/s0165-6147(99)01301-2

[101] Busija DW, Lacza Z, Rajapakse N, Shimizu K, Kis B, Bari F, et al. Targeting mitochondrial ATP-sensitive potassium channels—a novel approach to neuroprotection. Brain Research Reviews. 2004;**46**(3):282-294

[102] O'Rourke B. Evidence for mitochondrial K+ channels and their role in cardioprotection. Circulation Research. 2004;**94**(4):420-432

[103] Inoue I, Nagase H, Kishi K, Higuti T. ATP-sensitive K+ channel in the mitochondrial inner membrane. Nature. 1991;**352**(6332):244-247. DOI: 10.1038/352244a0

[104] Siemen D, Loupatatzis C, Borecky J, Gulbins E, Lang F. Ca2+−activated K channel of the BK-type in the inner mitochondrial membrane of a human glioma cell line. Biochemical and Biophysical Research Communications. 1999;**257**(2):549-554. DOI: 10.1006/bbrc.1999.0496

[105] Rusznák Z, Bakondi G, Kosztka L, Pocsai K, Dienes B, Fodor J, et al. Mitochondrial expression of the two-pore domain TASK-3 channels in malignantly transformed and non-malignant human cells. Virchows Archiv. 2008;**452**(4):415-426

[106] Szabò I, Bock J, Jekle A, Soddemann M, Adams C, Lang F, et al. A novel potassium channel in lymphocyte mitochondria. Journal of Biological Chemistry. 2005;**280**(13):12790-12798

[107] Bajgar R, Seetharaman S, Kowaltowski AJ, Garlid KD, Paucek P. Identification and properties of a novel intracellular (mitochondrial) ATP-sensitive potassium channel in brain. The Journal of Biological Chemistry. 2001;**276**(36):33369-33374. DOI: 10.1074/jbc.M103320200

[108] Dębska G, May R, Kicińska A, Szewczyk A, Elger CE, Kunz WS. Potassium channel openers depolarize hippocampal mitochondria. Brain Research. 2001;**892**(1):42-50

[109] Paggio A, Checchetto V, Campo A, Menabo R, Di Marco G, Di Lisa F, et al. Identification of an ATP-sensitive potassium channel in mitochondria. Nature. 2019;**572**:609-613

[110] Inagaki N, Gonoi T, JPt C, Namba N, Inazawa J, Gonzalez G, et al. Reconstitution of IKATP: An inward rectifier subunit plus the sulfonylurea receptor. Science. 1995;**270**(5239):1166-1170. DOI: 10.1126/science.270.5239.1166

[111] Waza AA, Andrabi K, Hussain MU. Protein kinase C (PKC) mediated interaction between conexin43 (Cx43) and K(+)(ATP) channel subunit (Kir6.1) in cardiomyocyte mitochondria: Implications in cytoprotection against hypoxia induced cell apoptosis. Cellular Signalling. 2014;**26**:1909-1917

[112] Duan P, Wang J, Li Y, Wei S, Su F, Zhang S, et al. Opening of mitoKATP

improves cardiac function and inhibits apoptosis via the AKT-Foxo1 signaling pathway in diabetic cardiomyopathy. International Journal of Molecular Medicine. 2018;**42**(5):2709-2719. DOI: 10.3892/ijmm.2018.3832

[113] Costa AD, Quinlan CL, Andrukhiv A, West IC, Jaburek M, Garlid KD. The direct physiological effects of mitoKATP opening on heart mitochondria. American Journal of Physiology-Heart and Circulatory Physiology. 2006;**290**(1):H406-HH15

[114] Köhler M, Hirschberg B, Bond C, Kinzie JM, Marrion N, Maylie J, et al. Small-conductance, calcium-activated potassium channels from mammalian brain. Science. 1996;**273**(5282):1709-1714

[115] Xu W, Liu Y, Wang S, McDonald T, Van Eyk JE, Sidor A, et al. Cytoprotective role of Ca2+− activated K+ channels in the cardiac inner mitochondrial membrane. Science. 2002;**298**(5595):1029-1033. DOI: 10.1126/science.1074360

[116] Singh H, Lu R, Bopassa JC, Meredith AL, Stefani E, Toro L. mitoBKCa is encoded by the Kcnma1 gene, and a splicing sequence defines its mitochondrial location. Proceedings of the National Academy of Sciences of the United States of America. 2013;**110**(26):10836-10841

[117] Stowe DF, Aldakkak M, Camara AK, Riess ML, Heinen A, Varadarajan SG, et al. Cardiac mitochondrial preconditioning by Big Ca2+−sensitive K+ channel opening requires superoxide radical generation. American Journal of Physiology. Heart and Circulatory Physiology. 2006;**290**(1):H434-H440. DOI: 10.1152/ajpheart.00763.2005

[118] Balderas E, Zhang J, Stefani E, Toro L. Mitochondrial BKCa channel. Frontiers in Physiology. 2015;**6**:104. DOI: 10.3389/fphys.2015.00104

[119] Saleh SN, Angermann JE, Sones WR, Leblanc N, Greenwood IA. Stimulation of Ca2+−gated Cl- currents by the calcium-dependent K+ channel modulators NS1619 [1,3-dihydro-1-[2-hydroxy-5-(trifluoromethyl) phenyl]-5-(trifluoromethyl)-2H-benzi midazol-2-one] and isopimaric acid. The Journal of Pharmacology and Experimental Therapeutics. 2007;**321**(3):1075-1084. DOI: 10.1124/jpet.106.118786

[120] Park WS, Kang SH, Son YK, Kim N, Ko JH, Kim HK, et al. The mitochondrial Ca2+−activated K+ channel activator, NS 1619 inhibits L-type Ca2+ channels in rat ventricular myocytes. Biochemical and Biophysical Research Communications. 2007;**362**(1):31-36. DOI: 10.1016/j.bbrc.2007.07.057

[121] Olesen S-P, Munch E, Moldt P, Drejer J. Selective activation of Ca2+−dependent K+ channels by novel benzimidazolone. European Journal of Pharmacology. 1994;**251**(1):53-59

[122] Holland M, Langton PD, Standen NB, Boyle JP. Effects of the BKCa channel activator, NS1619, on rat cerebral artery smooth muscle. British Journal of Pharmacology. 1996;**117**(1):119-129. DOI: 10.1111/j.1476-5381.1996.tb15163.x

[123] Frankenreiter S, Bednarczyk P, Kniess A, Bork NI, Straubinger J, Koprowski P, et al. cGMP-elevating compounds and ischemic conditioning provide cardioprotection against ischemia and reperfusion injury via cardiomyocyte-specific BK channels. Circulation. 2017;**136**(24):2337-2355. DOI: 10.1161/CIRCULATIONAHA.117.028723

[124] Gutman GA, Chandy KG, Grissmer S, Lazdunski M, McKinnon D,

Pardo LA, et al. International union of pharmacology. LIII. Nomenclature and molecular relationships of voltage-gated potassium channels. Pharmacological Reviews. 2005;**57**(4):473-508. DOI: 10.1124/pr.57.4.10

[125] Vicente R, Escalada A, Villalonga N, Texido L, Roura-Ferrer M, Martin-Satue M, et al. Association of Kv1.5 and Kv1.3 contributes to the major voltage-dependent K+ channel in macrophages. The Journal of Biological Chemistry. 2006;**281**(49):37675-37685. DOI: 10.1074/jbc.M605617200

[126] Leanza L, Henry B, Sassi N, Zoratti M, Chandy KG, Gulbins E, et al. Inhibitors of mitochondrial Kv1.3 channels induce Bax/Bak-independent death of cancer cells. EMBO Molecular Medicine. 2012;**4**(7):577-593. DOI: 10.1002/emmm.201200235

[127] Gulbins E, Sassi N, Grassme H, Zoratti M, Szabo I. Role of Kv1.3 mitochondrial potassium channel in apoptotic signalling in lymphocytes. Biochimica et Biophysica Acta. 2010;**1797**(6-7):1251-1259. DOI: 10.1016/j.bbabio.2010.01.018

[128] Szabó I, Bock J, Grassmé H, Soddemann M, Wilker B, Lang F, et al. Mitochondrial potassium channel Kv1. 3 mediates Bax-induced apoptosis in lymphocytes. Proceedings of the National Academy of Sciences of the United States of America. 2008;**105**(39):14861-14866

[129] Leanza L, Zoratti M, Gulbins E, Szabo I. Induction of apoptosis in macrophages via Kv1.3 and Kv1.5 potassium channels. Current Medicinal Chemistry. 2012;**19**(31):5394-5404. DOI: 10.2174/092986712803833281

[130] Leanza L, Biasutto L, Managò A, Gulbins E, Zoratti M, Szabo I. Intracellular ion channels and cancer. Frontiers in Physiology. 2013;**4**:227. DOI: 10.3389/fphys.2013.00227

[131] Toczyłowska-Mamińska R, Olszewska A, Laskowski M, Bednarczyk P, Skowronek K, Szewczyk A. Potassium channel in the mitochondria of human keratinocytes. Journal of Investigative Dermatology. 2014; **134**(3):764-772

[132] Enyedi P, Czirjak G. Molecular background of leak K+ currents: Two-pore domain potassium channels. Physiological Reviews. 2010;**90**(2):559-605. DOI: 10.1152/physrev.00029.2009

[133] Nagy D, Gonczi M, Dienes B, Szoor A, Fodor J, Nagy Z, et al. Silencing the KCNK9 potassium channel (TASK-3) gene disturbs mitochondrial function, causes mitochondrial depolarization, and induces apoptosis of human melanoma cells. Archives of Dermatological Research. 2014;**306**(10):885-902. DOI: 10.1007/s00403-014-1511-5

[134] Bernardi P, Petronilli V. The permeability transition pore as a mitochondrial calcium release channel: A critical appraisal. Journal of Bioenergetics and Biomembranes. 1996;**28**(2):131-138

[135] Di Lisa F, Carpi A, Giorgio V, Bernardi P. The mitochondrial permeability transition pore and cyclophilin D in cardioprotection. Biochimica et Biophysica Acta. 2011;**1813**(7):1316-1322. DOI: 10.1016/j.bbamcr.2011.01.031

[136] Elrod JW, Wong R, Mishra S, Vagnozzi RJ, Sakthievel B, Goonasekera SA, et al. Cyclophilin D controls mitochondrial pore–dependent Ca 2+ exchange, metabolic flexibility, and propensity for heart failure in

mice. Journal of Clinical Investigation. 2010;**120**(10):3680-3687

[137] Palty R, Silverman WF, Hershfinkel M, Caporale T, Sensi SL, Parnis J, et al. NCLX is an essential component of mitochondrial Na+/Ca2+ exchange. Proceedings of the National Academy of Sciences of the United States of America. 2010;**107**(1):436-441

[138] Sun J, Nguyen T, Aponte AM, Menazza S, Kohr MJ, Roth DM, et al. Ischaemic preconditioning preferentially increases protein S-nitrosylation in subsarcolemmal mitochondria. Cardiovascular Research. 2015;**106**(2): 227-236. DOI: 10.1093/cvr/cvv044

[139] Dang X, Doble BW, Kardami E. The carboxy-tail of connexin-43 localizes to the nucleus and inhibits cell growth. Molecular and Cellular Biochemistry. 2003;**242**(1-2):35-38

[140] Martins-Marques T, Ribeiro-Rodrigues T, Batista-Almeida D, Aasen T, Kwak BR, Girao H. Biological functions of connexin43 beyond intercellular communication. Trends in Cell Biology. 2019;**29**(10):835-847. DOI: 10.1016/j.tcb.2019.07.001

[141] Yang F, Chen WL, Zheng MZ, Yu GW, Xu HJ, Shen YL, et al. Heat shock protein 90 mediates anti-apoptotic effect of diazoxide by preventing the cleavage of Bid in hypothermic preservation rat hearts. The Journal of Heart and Lung Transplantation. 2011;**30**(8):928-934. DOI: 10.1016/j.healun.2011.04.001

[142] Garlid KD, Paucek P. Mitochondrial potassium transport: the K(+) cycle. Biochimica et Biophysica Acta. 2003;**1606**(1-3):23-41. DOI: 10.1016/s0005-2728(03)00108-7

[143] Augustynek B, Kunz WS, Szewczyk A. Guide to the pharmacology of mitochondrial potassium channels. In: Singh H., Sheu SS. (eds) Pharmacology of Mitochondria. Handbook of Experimental Pharmacology. Vol 240. Switzerland: Springer, Cham; 2016:103-127

[144] Leanza L, Checchetto V, Biasutto L, Rossa A, Costa R, Bachmann M, et al. Pharmacological modulation of mitochondrial ion channels. British Journal of Pharmacology. 2019;**176**(22):4258-4283. DOI: 10.1111/bph.14544

Section 2

Many Aspects of One Ion Transporter

Chapter 3

Hot on the Trail of Skin Inflammation: Focus on TRPV1/TRPV3 Channels in Psoriasis

Lisa S. Martin, Emma Fraillon, Fabien P. Chevalier and Bérengère Fromy

Abstract

Transient Receptor Potential Vanilloid (TRPV) channels are expressed in various skin cells, including non-neuronal cell types such as epidermal keratinocytes. They are polymodal sensors of the environment, regulating physiological function in response to a wide variety of stimuli. Indeed, in addition to their significant role in thermal responses and thermoregulation, TRPV channels are also implicated in local skin inflammation processes. Thus, these calcium permeable channels are associated to multiples skin diseases with inflammation, such as atopic dermatitis or psoriasis. In this chapter, we will mainly focus on TRPV1 and TRPV3 channels, as emerging pivotal targets for maintaining skin homeostasis in psoriasis-related inflammation.

Keywords: skin, epidermis, TRPV1, TRPV3, calcium channel, inflammation, psoriasis

1. Introduction

Skin is the largest organ of human organism, approximately 2m^2. This envelop, in constant contact with the environment, can be divided into three layers: a deep layer, the hypodermis; then an intermediate layer, the dermis; and finally, a superficial layer, the epidermis. This top layer is mainly constituted of keratinocytes, which form the first physical and chemical barrier between the external environment and our body. To maintain this function, keratinocytes undergo a multistep process of differentiation, from proliferating cells of the *stratum basale* to *stratum spinosum* (mature basal cells linked by keratin filaments – desmosomes), *granulosum* (mature keratinocytes, which generate keratin and keratohyalin granules), and *lucidum* (dead and flattened cells), to finally generate dead cornified corneocytes found in *stratum corneum* [1, 2]. These highly differentiated cells, devoid of nucleus and organelles, form the cornified envelop and are essential for the skin barrier function.

Despite its major keratinocyte content, epidermis is also composed of other cell populations in order to ensure protection of our organisms [1, 2]. Its immunity is guaranteed by Langerhans cells, a dendritic cell that contributes to innate and adaptative immunity [1, 2]. The sensory nerve endings passing through the epidermis (C-fibers and Aδ-fibers) were thought to be the exclusive transducers for the detection of environmental factors such as heat and pain, but Merkel cells can also act as mechanosensors [3]. Free intra-epidermal sensory nerve endings are all unmyelinated: C-fibers are unmyelinated, while Aδ-fibers lose their myelination when they enter the epidermis, allowing them to come into direct contact with the epidermal keratinocytes (**Figure 1A**) [4]. Therefore, in addition to the intra-epidermal sensory nerve endings, the epidermal keratinocytes also function as a sensory hub, able to detect environmental changes [5, 6]. Finally, the epidermis is constituted by another cell population: the melanocytes, located in the basal layer. With one for 4–10 keratinocytes, melanocytes provide a barrier from ultraviolet (UV) thanks to their ability to produce melanin, a photoprotector pigment [1, 2]. Opposite to the epidermis, the

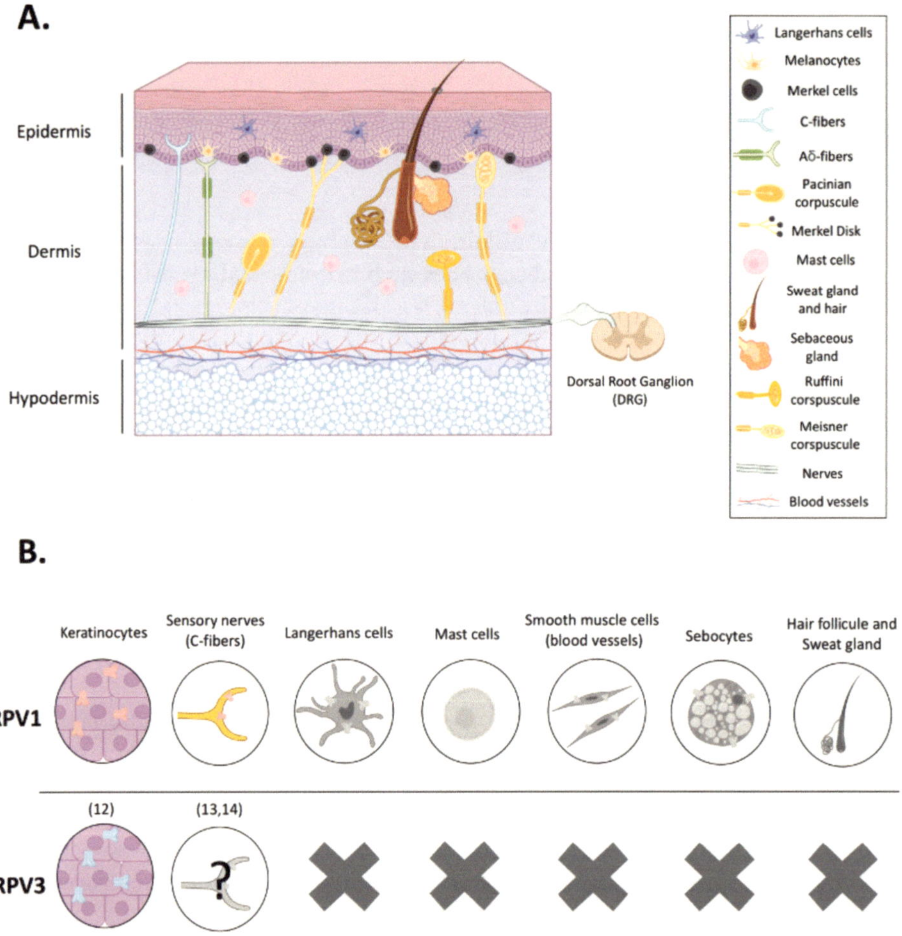

Figure 1.
Schematic representation of skin and TRPV1/TRPV3 location. A. Skin structure. Skin and its three layers: epidermis, dermis, and hypodermis. Nerve fibers are present in the epidermis or in the dermis depending on their properties and their type. Two major groups of skin nerve fibers are represented: Aδ fibers (green) poorly myelinated and able to pass through the dermo-epidermal junction, and Aβ fibers (yellow) strongly myelinated and not able to reach the epidermis. C-fibers are unmyelinated and able to pass through the dermo-epidermal junction. B. TRPV1 and TRPV3 location. TRPV1 is expressed in various cell types with a dominance in keratinocytes and sensory nerves (C fibers). TRPV3 expression is restricted to keratinocytes with a putative expression on sensory nerves.

dermis is mainly composed of extracellular matrix produced by dermal fibroblasts. This intermediate layer also supports dermal blood vessels, nerve fibers, and epidermal appendages (pilosebaceous unit and sweat glands). Finally, hypodermis is composed of adipocytes separated by connective tissue. This deep layer insulates and protects the skin from mechanical injuries [1, 2]. Thus, the skin allows a protection against externals insults (ultraviolet, pathogens, mechanical pressure, etc....) but also contributes to the maintenance of homeostasis such as information transfer, vitamin and metabolites secretions, hydric and thermal regulation.

To cope with various externals constrains, various cells of the skin express transmembrane sensors called Transient Receptor Potential (TRP) channels, which are involved in thermosensation, chemosensation, nociception, and mechanosensation [7]. TRP channels can be divided into six subfamilies: TRPA (ankyrin), TRPC (canonical), TRPM (melastatin), TRPML (mucolipin), TRPP (polycystin), and TRPV (vanilloid). Although TRPV channels have a major role in thermosensation [8], they also contribute to keratinocyte differentiation, skin barrier formation, and permeability thanks to calcium regulation. However, it appears that an aberrant TRP channel expression and function might contribute to some skin inflammatory diseases. In this review, we will focus on TRPV1 and TRPV3 structures, activation mechanisms, and their physiological roles in skin. We will also provide a new approach to study TRPV1 and TRPV3 channels in a very common chronic inflammatory disease, such as psoriasis.

2. TRPV1 and TRPV3 structure and gating

2.1 Expression and genetics

Among the six members of TRPV channels, thermosensors TRPV1 and TRPV3 are calcium channel both highly expressed in the skin, with different cells types expression (**Figure 1B**). The TRPV1 channel was firstly described on nociceptive sensory nerves from Dorsal Root sensory Ganglia (DRG) by Caterina *et al* in 1997 [9]. TRPV1 was detected on a subset of skin sensory nerves, such as peptidergic and non-peptidergic C fibers. Different studies have also proposed TRPV1 channel to be expressed on non-neuronal skin cells population. Indeed TRPV1 is expressed in human and mouse skin as TRPV1 immunoreactivity has been observed on Langerhans cells, mast cells, endothelium, and smooth muscle cells from dermal blood vessels, differentiated sebocytes, sweat glands, hair follicles (inner root and infundibulum), and finally on keratinocytes [10–12].

Unlike TRPV1, TRPV3 channels tissue expression is more restricted. Peier *et al.* (2002) have demonstrated the expression of Trpv3 in the skin, with a strong immunodetection on keratinocytes from epidermis and hair of rat [13]. We also confirmed a higher *TRPV3* expression in cultured human primary keratinocytes as compared with *TRPV1* (5.5-fold change, unpublished data). In contrast, *TRPV3* expression and activity on sensory nerves are still controversial. Indeed, *TRPV3* mRNA was detected on sensory neuron in DRG and trigeminal ganglia of monkey [14], while others have reported an absence of Trpv3 activity on mouse DRG, and then suggested no expression on these cells [15]. Finally, another group has proposed a heterodimeric form TRPV1-TRPV3 on sensory neurons [16]. Even if TRPV3 expression on sensory nerves is not completely clear, it is believed that epidermal TRPV3 and sensory nerves are able to communicate via chemical mediators. Indeed, TRPV3 activation in keratinocytes causes the secretion of an array of signaling factors, such as Nerve Growth Factor (NGF), Nitric Oxide (NO), Prostaglandin E2 (PGE_2), and Adenosine-Triphosphate (ATP).

TRPV1 (chr17:3,565,446-3,609,411) and TRPV3 (chr17:3,513,190-3,557,805) genes exist in tandem on human chromosome 17, with the same transcriptional orientation and are distant from less than 7650 base pairs, indicative of an ancestral gene duplication. In humans, the TRPV1 gene spans 17 exons encoding an 839 amino acids (aa) protein. Alternative splicing may occur and give rise to a modified amino acids sequence in the first 150 residues. The TRPV3 gene spans 18 exons encoding a prevalent isoform of 790 amino acids. As for TRPV1, TRPV3 might be differentially spliced, yielding two additional isoforms of 791 (additional A in position 760) and 765 amino acids (peptide sequence at 760–765 modified from DFNKIQ to GTVAVR together with deletion of residues 766–790). The most prevalent forms of TRPV3 (790 aa) and TRPV1 (839 aa) share 43% sequence homology.

2.2 Common features

The TRP superfamily is the second largest class of ions channels with a voltage-dependent activation mechanism. However, TRP members not only respond to electric signal but are also able to sense several environmental stimuli, rendering them polymodal sensors of the environment. These channels share a highly conserved protein architecture and require a tetrameric assembly to generate a functional central cation permeation pore. Apart from this ion channel pore, different subdomains of the TRPV proteins are responsible for their ability to be responsive to various environmental signals. Each subunit of the tetrameric complex is composed of six transmembrane

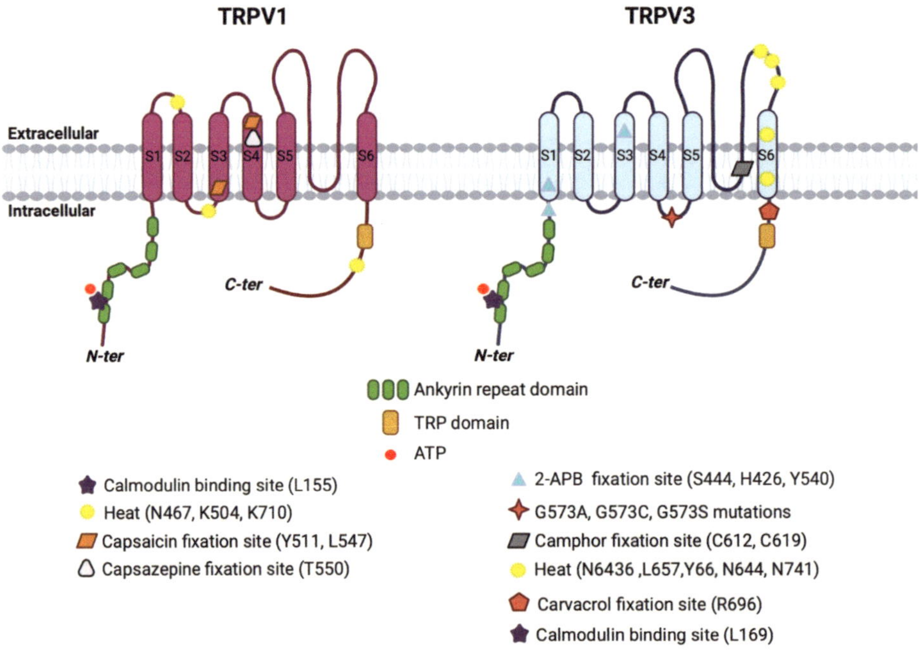

Figure 2.
TRPV1 and TRPV3 structure. Left – Highlighting the fixation sites of TRPV1 activators (heat, capsaicin) and interacting protein (calmodulin). Right – Fixation sites of TRPV3 activators (2-APB, carvacrol, camphor, heat), interacting protein (calmodulin) and mutations involved in the establishment of the inflammatory response (G573A, G573C, G573S). Ankyrin repeat domain and TRP domain are conserved in TRPV1 and TRPV3 structures.

segments (S1–S6), with the cation permeable pore formed by a reentrant loop located between S5 and S6 (**Figure 2**). The S1–S4 bundle likely forms the voltage sensor module, although TRPV1 and TRPV3 exhibit a weak voltage dependence for gating. This relatively low capacity of TRPV channels to respond to electric current may be due to the scarcity of positively charged amino acids in the S4 domain that contains a single arginine residue [17, 18]. The large cytoplasmic N-terminal part of a monomer comprises six ankyrin repeats, each consisting of a 33-residue motif forming two antiparallel alpha helices separated by loops linking the adjacent repeats. The ankyrin repeat domains (ARDs) are highly conserved in TRPV1 and TRPV3. The C-terminal part is also a large intracellular region containing a TRP-domain, another highly conserved distinctive and fundamental feature of TRP channels, consisting of a 25 amino acids α-helix structure with a conserved sequence IWKLQR called the TRP box. This TRP domain is running parallel to the inner plasma membrane and intimately lodged within the intracellular side of the S1–S4 module [19, 20]. A coiled-coil motif comprising residues of E684–R721 and overlapping with the TRP domain in the C-terminus of TRPV1 has been identified as an association domain and appears to be the molecular determinant of tetramerization [21]. In TRPV3, the C-terminal domain additionally forms a loop of 19 residues from V737 to V756, which has not been observed in other TRPV channels. This unique C-terminal loop domain extends along the TRPV3 intracellular skirt and forms an extensive network of interactions with ankyrin repeats 2–5 of the ARD [22]. Among all of these domains, several regions and specific residues have been mapped within TRPV1 and TRPV3 to regulate the channel gating. Indeed, heat, voltage, and ligand stimuli are sensed by TRPV channels by different structural domains.

2.3 Pore module

Both pores of TRPV1 and TRPV3 are prone to dilatation during stimulation, rendering them permeable for large cations [19, 23]. Thus, TRPV1 and TRPV3 are cation-selective channels exhibiting a notable preference for divalent cations, with the following permeability sequence: Ca2+ > Mg2+ > Na + ≈K+. As for all TRPV channels, negatively charged amino acids in the pore play a central role in cation permeation, and furthermore, the opened pore is blocked by both extra- and intracellular cations. The S5-S6 segments are forming the central pore and the lower gate. This lower gate is formed by a hydrophobic seal, blocking permeation by hydrated ions when the channels are in their closed state. This hydrophobic seal is ensured by the critical residue I679 on S6 for TRPV1 and M677 on S6 for TRPV3 [20, 22]. An additional upper gate is formed by a short loop and helix between S5 and S6, called the pore helix (PH), and acts as a selectivity filter. TRPV1 displays a prolonged loop of 23 residues between S5 and the PH, named the pore turret. This pore turret is a mandatory structural domain for conformational rearrangements during heat activation of the TRPV1 channel, but is not part of the capsaicin agonist activation pathway [24, 25]. In TRPV3, the three key amino acids I644, N647, and Y661 located in the S6 are responsible for heat activation of the channel, since single-point mutants of these generate total loss of temperature activation, without affecting the overall TRPV3 structure [26]. Interestingly, the temperature sensitivity of the TRPV1 channel also implies the C-terminal domain [27].

2.4 Ligands

TRPV1 can be activated by numerous exogenous agonists including capsaicin, plant toxin resiniferatoxin (RTX), natural substances such as capsaicin-related compounds from peppers, aromatic components, and animal vanillotoxins from the venom of the tarantula [28, 29]. For TRPV1, the three amino acids R491, Y511, and S512 in the S3 transmembrane segment are responsible for capsaicin sensitivity, while the region between S481 and T550 is responsible for binding of the antagonist capsazepine, without affecting the temperature activation [30, 31]. Natural substances also activate TRPV3, including camphor (C612 and C619), carvacrol, thymol, and eugenol [32]. Moreover, TRPV3 can be activated by synthetic molecules such as the well-documented 2-Aminoethoxydiphenyl Borate (2-APB).

2-APB is a common activator ligand of TRPV1 and TRPV3 channels [27]. In TRPV3, several transmembrane segments are implicated in the binding of the agonist 2-APB. In fact, there are three different sites of 2-APB fixation in TRPV3. A first site of fixation is involving S444 of S1, E501 and W493 of S2, and Y565, H523, and F526 of S3 that establish complementary interactions with different atoms of 2-APB [22]. A second site of 2-APB binding is mostly mediated by polar residues, such as H417 and T421 of the linker domain, H426 and H430 of the pre-S1 helix, and R693 and R696 of the TRP domain. Interestingly, the mutation H426A completely abolishes TRPV3 activation by 2-APB but not by camphor neither by carvacrol [22, 33, 34]. However, the residue R696 in the TRP domain appears critical in TRVP3 activation by external ligands since the mutation R696K abolishes 2-APB- and carvacrol-induced calcium influx [34]. The third site of 2-APB fixation is nested in a cavity formed by the extracellular portions of helices S1-S4 and is mediated through both hydrophobic and hydrophilic residues including V458, Y540, R487, and Q483 [22]. The binding of 2-APB on the first two sites described above does not induce gating-associated conformational changes. In opposite, dramatic structural rearrangements are observed when 2-APB binds the third site [22]. Thus, binding of 2-APB to the first two sites is likely a prerequisite for gating, by stabilizing the multiple domains during channel opening.

Endogenous ligands were also reported for TRPV1 and TRPV3: unsaturated N-acyldopamines, lipoxygenase products of arachidonic acid, linoleic acid, Phospholipase C metabolites, and the endocannabinoid anandamide [35].

2.5 Sensitization/desensitization

In both TRPV1 and TRPV3, the N-terminus module contains a domain able to bind calmodulin (CaM) in a Ca^{2+}-dependent manner [36]. This domain is located between the ankyrin repeats 2 and 3, which comprise a conserved site (K155 and K160 for TRPV1; K169 and K174 for TRPV3) involved in both CaM and ATP binding [37, 38]. In a resting cell, ATP is bound, and the channel is sensitized. Indeed, it has been shown that ATP binding to the TRPV1-ARD generates larger currents in response to capsaicin application [39, 40]. Hence, after channel opening, Ca^{2+} flows inward and chelates the ATP, which is released from TRPV1-ARD, thus freeing the binding site. In parallel, the Ca^{2+} influx activates CaM, and Ca^{2+}-CaM can replace the sensitizer and engages the ARD to close the channel [40]. Thus, CaM is involved in Ca^{2+}-dependent desensitization of TRPV1 [41].

The binding of ATP and Ca^{2+}-CaM to the N-terminal ARD observed in TRPV1 is conserved in TRPV3 [38], although differences exist. TRPV1 is desensitized after cumulative stimulations. In contrast, TRPV3 is the only member among the TRP channels that sensitizes upon repeated application of stimuli. In addition, the sensitization of TRPV3 is independent of the origin of the stimulus, it will sensitize regardless whether it is activated by heat or chemical ligands. TRPV3 also displays cross-sensitization to stimuli of a different nature, as camphor stimulation causes a sensitization to heat [42]. It is known that sensitization is due to the decrease of the inhibition by calcium from both sides of cells [43]. In the intracellular side, Ca^{2+}-CaM binds ARD and inhibits TRPV3, as described above for TRPV1. The TRPV3-ARD structure is very close to ARD from other members of TRPV channel family, except it exhibits a unique particular conformation of finger 3 loop. This linker region in between the ankyrin repeats 3 and 4 is greatly stabilized by a network of hydrogen bonds and an hydrophobic environment, instead of being flexible as seen in the other TRPV-ARD arrangements [44]. This stabilized finger 3 of TRPV3-ARD may cause steric hindrance, which impedes the binding of CaM. Therefore, CaM binding to ARD probably forces a conformational change of finger 3, thus resulting in an inhibition of TRPV3 function. Upon successive simulations, the finger 3 of TRPV3-ARD undergoes conformational change that decreases the binding of CaM, causing the channel to open more easily. Thus, the finger 3 of TRPV3-ARD functions as a switch in regulation of TRPV3 upon stimulation. Moreover, this distinctive finger 3 segment precedes a conserved threonine 264, which has been identified as a putative site for the ERK1-dependent modulation of TRPV3 [45]. Phosphorylation events could, therefore, alter the conformation of this important loop and powerfully influence the binding of regulatory factors [46]. In others contexts, the influence of phosphorylation events has been demonstrated especially for TRPV1, where phosphorylation of the channel induces sensitization, whereas dephosphorylation is associated to desensitization [47]. The TRPV1 C-terminus also contains modulatory domains able to be phosphorylated and to bind CaM through the 35 amino acids segment E767–T801 [41, 48].

In contrast to TRPV1, the naive TRPV3 channel does not show any intrinsic voltage-dependent activation. The voltage dependence only appears when the TRPV3 channel is primo-stimulated by chemicals or heat stimulus [43, 49]. This voltage dependence of TRPV3 is established by Ca^{2+} binding on N641 at the pore loop after opening. In addition, the voltage dependence is strongly influenced by Ca^{2+}-CaM binding at the cytoplasmic N terminus. Sensitization is accompanied by a decrease in the voltage dependence. Finally, the sensitized TRPV3 channels are less inhibited than the naive ones, showing faster activation at positive potentials and less deactivation at negative potentials. This gradual shift in Ca^{2+}-dependent regulation or TRPV3 activity is likely related to conformational changes after successive stimulations [43, 44, 46]. Considering the huge complexity in the structural arrangement and interactions of the multiple domains of the TRPV channels, it has been difficult to fully decipher the mechanisms of gating, and many questions remain open.

3. TRPV1 and TRPV3 channels in skin function

3.1 Epidermal barrier function

Ca^{2+} is well known to contribute to epidermal homeostasis and thus to the formation of an effective skin barrier [50, 51]. In order to maintain its barrier function, the epidermis needs to be renewed every 28 days depending on a calcium gradient. The increase in calcium concentration in the outer layer is essential for the terminal differentiation of keratinocytes, which will lead to the formation of the *stratum corneum* and ensure the skin's physical barrier role [51]. This supports the role of calcium-permeable channels in epidermal barrier function.

The role of TRPV3 in the epidermal differentiation process was highlighted after aberrant expression of early differentiation markers (i.e., KRT1/KRT10) in keratinocytes from

Trpv3-KO mice [52]. In addition, the decrease in transglutaminase activity contributed to an alteration in the *stratum corneum* formation. This regulation of keratinocyte differentiation process appears to be dependent on the TRPV3/TGFα/EGFR signaling axis [52, 53]. These data support the importance of TRPV3 channels as actors in the balance between proliferation and differentiation, thus giving them a crucial role in skin barrier formation. In contrast, the contribution of TRPV1 channels in the skin barrier remains unknown.

3.2 Sensory modalities in the healthy skin

TRPV1 is a major nonselective cation channel with polymodal mechanisms of activation [54]. Functional TRPV1 serves as a thermal sensor since it is gated by noxious heat greater than 42°C and also chili pepper [9, 55]. The heat nociception was almost abolished following ablation of TRPV1-expressing neurons in mice [56], but *Trpv1*-knockout mice display only a partial defect in the ability to sense and respond to acute noxious heat [57]. Partial explanation could be that three channels, including TRPV1, TRPM3, and TRAP1, act in concert to mediate behavioral responses to noxious heat [58]. Besides heat, TRPV1 channels can be directly activated by proton, such as extracellular acidification (pH less than ~6.0) [9, 59]. Consistently, an integrative study performed in healthy rats has shown that cutaneous vasodilation in response to cathodal stimulation was induced by TRPV1 channels, likely through local acidification and the PGIS/PGI2/IP pathway [60].

In contrast to TRPV1, TRPV3 was reported as a warm sensor for innocuous temperatures (approximately 33°C) in different *in vitro* studies [13, 14, 16]. Not surprisingly, a profound deficit in sensing warm external temperatures was described in mice lacking Trpv3 [15]. Later it has been shown that *Trpv3*-KO mice displayed a preference toward cooler temperatures [61], showing that TRPV3 influences thermal information that is used to modulate thermal comfort or preference. TRPV3 on keratinocytes has been shown to play a role in thermosensation involving ATP signaling, but other molecules have also been reported [62]. The overexpression of TRPV3 in keratinocyte induces the release of PGE_2, an algogenic substance. Interestingly, in mice overexpressing Trpv3 channels selectively in keratinocytes, an hyperalgesia was observed [63]. These data support that TRPV3 channels participate to the thermal and pain transduction through these mediators. More recently, *in vivo* demonstration was provided for the role of cutaneous Trpv3 as a warm sensor of heating and a strong modulator of cutaneous vascular thermoregulatory mechanisms [64]. Since keratinocytes are representing the primary site of action of TRPV3 in mice (see the above section on TRPV3 expression), this study indicates that TRPV3 channels in the keratinocytes serve as heat detectors for warm temperatures to regulate cutaneous thermal homeostasis via initial changes in local blood flow. In contrast, TRPV1 channels are not involved in this process since *Trpv1*-KO mice displayed a normal heat-evoked vasodilation. It is interesting to note that *Trpv3*-KO mice showed a delay in behavioral response to noxious temperature over 50°C and 55°C, indicating that TRPV3 is also involved in response to acute painful heat stimuli [15]. This coincides with the ability to sensitize TRPV3 upon noxious heat stimuli [13]. Since *Trpv3*-KO mice and *Trpv1*-KO mice have an identical thermal nociceptive phenotype, this suggests that these two channels have overlapping thermal detection and functions.

4. TRPV1 and TRPV3 channels in psoriasis

Psoriasis is a chronic multifactorial inflammatory disease, resulting from the interaction between genetic predisposing factors and environmental triggers, with a global incidence ranging between 0.09% and 5.1% [65]. This dermatosis results from altered signaling between epidermal keratinocytes and the immune system leading to an uncontrolled keratinocyte proliferation (hyperplasia), impaired keratinocyte differentiation (hyperkeratosis), and chronic inflammation. The immune cells (i.e., dendritic and T cells) infiltrating skin lesions produce a wide variety of cytokines (IL-23, IL-17, IFNγ), which activate keratinocytes [66, 67]. Once activated, keratinocytes produce pro-inflammatory cytokines (i.e., IL-6, TNFα), chemokines (i.e., CXCL1, CCL2, CCL13), and antimicrobial peptides (cathelicidin, β-defensine) that further stimulate immune cells and thus maintain the disease in a chronic state. Therefore, keratinocytes not only respond to inflammation but also contribute to the recruitment and the activation of immune cells. Moreover, TRPV1 and TRPV3 channels have a predominant role in inflammation, pain, and pruritus and could be involved in the vicious cycle of the inflammation process in psoriasis.

Indeed, TRPV1 overexpression has been found in the skin of psoriatic patients and in the mouse model of "imiquimod-induced psoriasis" [68, 69]. Imiquimod (IMQ) is a potent immune activator stimulating the IL-23/IL-17 axis and thus mimicking psoriasis inflammation [70]. In 2014, Riol-Blanco *et al* have shown that ablation by RTX of $TRPV1^+/NaV1.8^+$ nociceptors reduces immune cells infiltration and psoriasis skin inflammation, by acting on IL-23 and IL17 release (**Figure 3**) [71]. Furthermore, this study revealed that $TRPV1^+/NaV1.8^+$ nociceptors can interact with dermal dendritic cells to regulate the IL-23/IL-17 axis during the initiation phase of psoriasis. Further data support a role for TRPV1 in this skin disease. In 2018, Zhou *et al.* also highlighted a significant decrease in epidermal hyperplasia, inflammatory cell infiltration, and cytokine production (IL-1, IL-6, IL-23) in IMQ-treated *Trpv1*-KO mice [69]. The NGF-TrkA-TRPV1 signaling pathway in nerve fibers has also been shown to play a role in psoriasis lesion formation [72]. Indeed, both nerve growth factor (NGF) and Tropomyosin receptor kinase A (TrkA) are highly expressed in psoriasis, and their interaction induces activation of TRPV1-mediated pain and pruritus [72–74]. Consistently, a TrkA kinase inhibitor (CT327) reduced pruritus of psoriatic patients (15). It is possible that NGF could sensitize TRPV1 channels on nerve fibers ($NaV1.8^+$) during the initiation phase and then stimulate innate immune dermal dendric cells for the induction of IL-23/IL-17 signaling, leading to the development of psoriasis. Together, these data confirm the involvement of TRPV1 in psoriasis, with a major role of sensory nerve fibers.

In contrast to TRPV1, TRPV3 channels seem to act on psoriasis from upper cell layers, such as keratinocytes. Indeed, TRPV3 channel expression is increased in psoriatic skin lesions, with significant labeling within the epidermis [68, 75]. Moreover, many studies have supported the role of TRPV3 on inflammation. Upon stimulation, TRPV3 can activate the EGFR/NFκB pathway and induce the release of mediators, such as IL-1α, IL-6, IL-8, TNFα, ATP, and PGE2 [53, 76], which will in turn act on sensory nerves and provoke pain and itching [62, 63]. In addition, Zhao *et al.* recently proposed that protease-activated receptor (PAR2) sensitizes TRPV3 channels on keratinocytes, resulting in secretion of thymic stromal lymphopoietin (TSLP), a potent pro-inflammatory cytokine. The latter then contributes to the production of IL-23 by dendritic cells and induces severe itching [77, 78]. Finally, the TRPV3 gain-of-function mutations G573S or G573C result in hyperkeratosis in mice, increased inflammatory cytokines in serum (IL-1α, Il-6, IL-17), and high levels of pruritogenic substances, such as NGF (**Figure 3**) [79, 80]. To further endorse the role of TRPV3 in inflammation, the 17(R)-resolvin D1, an anti-inflammatory lipid, is able to suppress TRPV3-induced hypersensitivity/pain during the inflammatory response

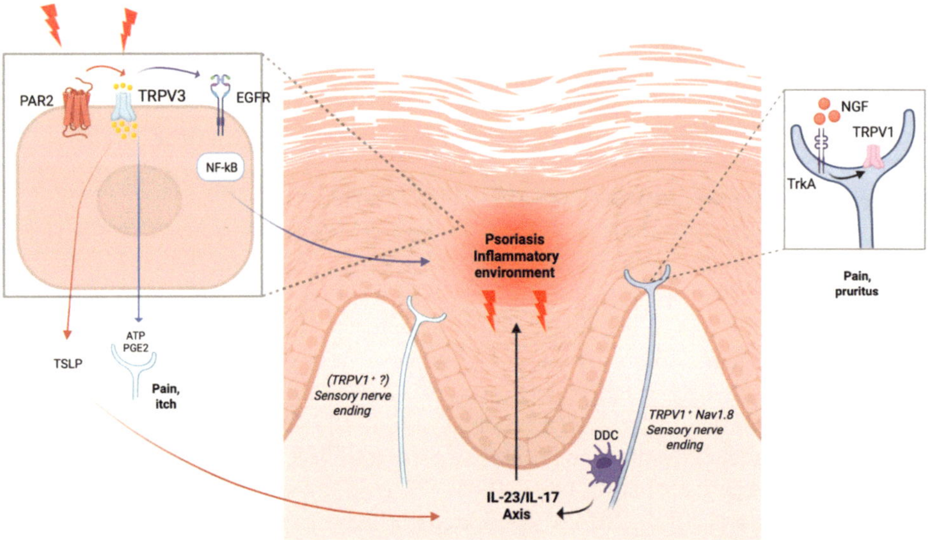

Figure 3.
TRPV1 and TRPV3 in the physiopathology of psoriasis. Left – Inflammatory environment in skin psoriasis stimulates PAR2 on keratinocytes. Activated PAR2 stimulates TRPV3, which leads to the production of TSLP, an activator of the IL-23/IL-17 axis (red arrows). Inflammation also directly activates TRPV3, which leads to the production and the fixation of ATP and PGE2 on C-fibers, leading to pain and itch (blue arrows). Activated TRPV3 also stimulates and activates EGFR. In this context, the EGFR activation induces the transcription of NF-κB, allowing the production of inflammatory cytokines that sustain the inflammatory process in the skin. Right – In a psoriasis skin, keratinocytes release NGF in the extracellular environment. NGF can interact with TrkA present on the TRPV1+/Nav1.8 nerve endings. This interaction activates TRPV1 and leads to pain and pruritus. The interaction of dermal dendritic cells (DDC) with TRPV1+/Nav1.8 nerve endings potentiates the activation of the IL-23/IL-17 axis.

in mice [81]. Moreover, injection of the TRPV3 antagonist 74a is able to attenuate pruritus and inflammatory response in mice with chronic inflammatory disease such as atopic dermatitis [77]. Thus, either through the activation of signaling pathways (EGFR/NFκB; PAR2) or the release of algogenic and pruritogenic mediators, TRPV3 channels appear to play a role in the initiation of psoriasis.

Altogether, these data support that TRPV1 and TRPV3 are actors of inflammation, either from sensory nerve fibers or from keratinocytes. Both channels seem to have fundamental role in the pathogenesis of psoriasis due to their ability to activate immune cells or to induce cytokine production. It could also be hypothesized that these two channels cooperate during the initiation process of psoriasis.

5. TRPV1 and TRPV3 channels in therapeutic perspectives of psoriasis

The well-known biological strategies to treat an inflammatory disease such as psoriasis consist of using a specific blocking antibody to freeze the interaction of a cytokine with its receptor and then blocking the underlying cytokine-specific biological response. As mentioned earlier, both TRPV1 and TRPV3 are involved in the inflammatory process of psoriasis and could also be potential therapeutic targets in this disease. Indeed, TrkA appears to be a potential target because its inhibition with CT327 blocks the overactivation of TRPV1 on sensory neurons. The discovery of this new molecule is promising since the decrease in pain and pruritus has been observed in psoriatic patients [72]. Interestingly, another study also demonstrated a decrease in skin hyperplasia and erythema in psoriasis murine models treated with the TRPV1 antagonist SB366791 [82]. Research on the inhibition of TRPV1 in psoriasis disease has already emerged and needs to be further explored.

As the TRPV3/TGFα/EGFR signaling complex seems to play a role in psoriasis, EGFR also becomes a potential therapeutic target. A recent clinical study showed that erlotinib, an EGFR inhibitor, significantly decreases hyperkeratosis and pain in Olmsted syndrome patients displaying a TRPV3 overactivation mutation [83]. This highlights that inhibition of EGFR could decrease keratinocytes hyperproliferation/differentiation and might also act on inflammation and pain. Since TRPV3 acts at the first line, directly targeting this channel could also be very promising. Indeed, we reported above some anti-inflammatory molecules, such as 17(R)-resolvin D1, which suppresses TRPV3-induced pain and inflammation. Another potential promising molecule is the TRPV3 antagonist called 74a, which has already proved its efficacy on another inflammatory skin disease [77].

It could be suggested that inflammatory environment activates TRPV3 from keratinocytes and TRPV1+ nerves for the induction of pain, itching, and inflammation. The cooperation of these two channels, with the potential activation of TRPV1 induced by downstream mediators of TRPV3, cannot be excluded. To increase further the complexity of the TRPV1 and TRPV3 relationship, it has been demonstrated that TRPV1 and TRPV3 could form interacting partners, therefore assembling heterochannels [84, 85]. The biological significance of these heterochannels is still not known, but this could be involved in a very fine-tuning of sensitivity. In addtion, a recent study showed that the intergenic region between TRPV1 and TRPV3 coding sequences contained a human specific transposable element (SVA: SINE-VNTR-Alu retrotransposon) insertion, located upstream of the TRPV3 promoter and downstream of the 3′ end of TRPV1 [86]. This SVA insertion acts as a cis-regulatory element allowing coexpression of TRPV1 and TRPV3 in multiple human tissues, which is not observed in mice. Thus, targeting these two channels simultaneously could be a very promising approach for the treatment of psoriasis in human.

6. Conclusion

TRPV1 and TRPV3 are important channels for maintaining the skin homeostasis and its function. A deregulation of their expression and/or activity is associated to the establishment of an inflammatory response. Several studies already demonstrated their contribution in psoriasis by highlighting their capacities to activate typical inflammatory pathways (IL-23/IL-17 axis, NF-κB pathway) and/or to provoke the release of inflammatory mediators. It would therefore be interesting to explore the exact contribution of TRPV1 and TRPV3 in the psoriasis typical chronic skin inflammation: are they the trigger or are they one of the multiple factors involved in maintaining inflammation?

It would be interesting to explore the presence of TRPV1 and TRPV3 in other tissues of the body subjected to chronic inflammation, such as the intestinal epithelium. The future challenge will be to develop specific compounds that target the TRPV1/TRPV3 to limit chronic inflammation without affecting their physiological functions.

Conflict of interest

The authors declare no conflict of interest.

Notes

All of the artworks used were adapted from the illustration bank of Biorender (https://biorender.com).

Author details

Lisa S. Martin[1,2], Emma Fraillon[1,2], Fabien P. Chevalier[1,2]* and Bérengère Fromy[1,2]*

1 CNRS UMR 5305, Tissue Biology and Therapeutic Engineering Laboratory, Lyon, France

2 Claude Bernard University Lyon 1, Villeurbanne, France

*Address all correspondence to: fabien.chevalier@univ-lyon1.fr
and berengere.fromy@univ-lyon1.fr

References

[1] Wong R, Geyer S, Weninger W, Guimberteau J-C, Wong JK. The dynamic anatomy and patterning of skin. Experimental Dermatology. 2016;**25**(2):92-98

[2] Khavkin J, Ellis DAF. Aging Skin: Histology, Physiology, and Pathology. Facial Plastic Surgery Clinics of North America. 2011;**19**(2):229-234

[3] Roosterman D, Goerge T, Schneider SW, Bunnett NW, Steinhoff M. Neuronal Control of Skin Function: The Skin as a Neuroimmunoendocrine Organ. Physiological Reviews. 2006;**86**(4):1309-1379

[4] Talagas M, Lebonvallet N, Leschiera R, Marcorelles P, Misery L. What about physical contacts between epidermal keratinocytes and sensory neurons? Experimental Dermatology. 2018;**27**(1):9-13

[5] Denda M, Nakatani M, Ikeyama K, Tsutsumi M, Denda S. Epidermal keratinocytes as the forefront of the sensory system. Experimental Dermatology. 2007;**16**(3):157-161

[6] Moehring F, Cowie AM, Menzel AD, Weyer AD, Grzybowski M, Arzua T, et al. Keratinocytes mediate innocuous and noxious touch via ATP-P2X4 signaling. eLife. 2018;7:e31684

[7] Venkatachalam K, Montell C. TRP Channels. Annual Review of Biochemistry. 2007;**76**:387-417

[8] Voets T. TRP channels and thermosensation. Handbook of Experimental Pharmacology. 2014;**223**:729-741

[9] Caterina MJ, Schumacher MA, Tominaga M, Rosen TA, Levine JD, Julius D. The capsaicin receptor: a heat-activated ion channel in the pain pathway. Nature. 1997;**389**(6653):816-824

[10] Bodó E, Kovács I, Telek A, Varga A, Paus R, Kovács L, et al. Vanilloid Receptor-1 (VR1) is Widely Expressed on Various Epithelial and Mesenchymal Cell Types of Human Skin. The Journal of Investigative Dermatology. 2004;**123**(2):410-413

[11] Ständer S, Moormann C, Schumacher M, Buddenkotte J, Artuc M, Shpacovitch V, et al. Expression of vanilloid receptor subtype 1 in cutaneous sensory nerve fibers, mast cells, and epithelial cells of appendage structures. Experimental Dermatology. 2004;**13**(3):129-139

[12] Denda M, Fuziwara S, Inoue K, Denda S, Akamatsu H, Tomitaka A, et al. Immunoreactivity of VR1 on epidermal keratinocyte of human skin. Biochemical and Biophysical Research Communications. 2001;**285**(5):1250-1252

[13] Peier AM. A Heat-Sensitive TRP Channel Expressed in Keratinocytes. Science. 2002;**296**(5575):2046-2049

[14] Xu H, Ramsey IS, Kotecha SA, Moran MM, Chong JA, Lawson D, et al. TRPV3 is a calcium-permeable temperature-sensitive cation channel. Nature. 2002;**418**(6894):181-186

[15] Moqrich A. Impaired Thermosensation in Mice Lacking TRPV3, a Heat and Camphor Sensor in the Skin. Science. 2005;**307**(5714):1468-1472

[16] Smith GD, Gunthorpe MJ, Kelsell RE, Hayes PD, Reilly P, Facer P, et al. TRPV3 is a temperature-sensitive

vanilloid receptor-like protein. Nature. 2002;**418**(6894):186-190

[17] Voets T, Droogmans G, Wissenbach U, Janssens A, Flockerzi V, Nilius B. The principle of temperature-dependent gating in cold- and heat-sensitive TRP channels. Nature. 2004;**430**(7001):748-754

[18] Nilius B, Talavera K, Owsianik G, Prenen J, Droogmans G, Voets T. Gating of TRP channels: a voltage connection? The Journal of Physiology. 2005;**567** (Pt 1):35-44

[19] Cao E. Structural mechanisms of transient receptor potential ion channels. Journal of General Physiology. 2020;**152**(3):e201811998

[20] Liao M, Cao E, Julius D, Cheng Y. Structure of the TRPV1 ion channel determined by electron cryo-microscopy. Nature. 2013;**504**(7478):107-112

[21] García-Sanz N, Fernández-Carvajal A, Morenilla-Palao C, Planells-Cases R, Fajardo-Sánchez E, Fernández-Ballester G, et al. Identification of a tetramerization domain in the C terminus of the vanilloid receptor. The Journal of Neuroscience. 2004;**24**(23):5307-5314

[22] Singh AK, McGoldrick LL, Sobolevsky AI. Structure and gating mechanism of the transient receptor potential channel TRPV3. Nature Structural & Molecular Biology. 2018;**25**(9):805-813

[23] Chung M-K, Güler AD, Caterina MJ. TRPV1 shows dynamic ionic selectivity during agonist stimulation. Nature Neuroscience. 2008;**11**(5):555-564

[24] Cui Y, Yang F, Cao X, Yarov-Yarovoy V, Wang K, Zheng J. Selective disruption of high sensitivity heat activation but not capsaicin activation of TRPV1 channels by pore turret mutations. Journal of General Physiology. 2012;**139**(4):273-283

[25] Yang F, Cui Y, Wang K, Zheng J. Thermosensitive TRP channel pore turret is part of the temperature activation pathway. Proceedings of the National Academy of Sciences. 2010;**107**(15):7083-7088

[26] Grandl J, Hu H, Bandell M, Bursulaya B, Schmidt M, Petrus M, et al. Pore region of TRPV3 ion channel is specifically required for heat activation. Nature Neuroscience. 2008;**11**(9):1007-1013

[27] Colton CK, Zhu MX. 2-Aminoethoxydiphenyl Borate as a Common Activator of TRPV1, TRPV2, and TRPV3 Channels. In: Flockerzi V, Nilius B, éditeurs. Transient Receptor Potential (TRP) Channels. Berlin, Heidelberg: Springer Berlin Heidelberg; 2007. p. 173-187. (Handbook of Experimental Pharmacology; vol. 179).

[28] Siemens J, Zhou S, Piskorowski R, Nikai T, Lumpkin EA, Basbaum AI, et al. Spider toxins activate the capsaicin receptor to produce inflammatory pain. Nature. 2006;**444**(7116):208-212

[29] Vriens J, Appendino G, Nilius B. Pharmacology of vanilloid transient receptor potential cation channels. Molecular Pharmacology. 2009;**75**(6):1262-1279

[30] Jordt S-E, Julius D. Molecular basis for species-specific sensitivity to hot chili peppers. Cell. 2002;**108**(3):421-430

[31] Marín-Burgin A, Reppenhagen S, Klusch A, Wendland JR, Petersen M. Low-threshold heat response antagonized by capsazepine in chick sensory neurons, which are capsaicin-insensitive. The

European Journal of Neuroscience. 2000;**12**(10):3560-3566

[32] Broad LM, Mogg AJ, Eberle E, Tolley M, Li DL, Knopp KL. TRPV3 in Drug Development. Pharmaceuticals (Basel). 2016;**9**(3):55

[33] Hu H, Grandl J, Bandell M, Petrus M, Patapoutian A. Two amino acid residues determine 2-APB sensitivity of the ion channels TRPV3 and TRPV4. Proceedings of the National Academy of Sciences of the United States of America. 2009;**106**(5):1626-1631

[34] Deering-Rice CE, Mitchell VK, Romero EG, Abdel Aziz MH, Ryskamp DA, Križaj D, et al. Drofenine: a 2-APB analog with improved selectivity for human TRPV3. Pharmacology Research & Perspectives. 2014;**2**(5)

[35] Van Der Stelt M, Di Marzo V, Endovanilloids. Putative endogenous ligands of transient receptor potential vanilloid 1 channels. European Journal of Biochemistry. 2004;**271**(10):1827-1834

[36] Rosenbaum T, Gordon-Shaag A, Munari M, Gordon SE. Ca2+/calmodulin modulates TRPV1 activation by capsaicin. The Journal of General Physiology. 2004;**123**(1):53-62

[37] Nilius B, Bíró T, Owsianik G. TRPV3: time to decipher a poorly understood family member!: TRPV3. The Journal of Physiology. 2014;**592**(2):295-304

[38] Phelps CB, Wang RR, Choo SS, Gaudet R. Differential Regulation of TRPV1, TRPV3, and TRPV4 Sensitivity through a Conserved Binding Site on the Ankyrin Repeat Domain. Journal of Biological Chemistry. 2010;**285**(1):731-740

[39] Lau S-Y, Procko E, Gaudet R. Distinct properties of Ca2+–calmodulin binding to N- and C-terminal regulatory regions of the TRPV1 channel. Journal of General Physiology. 2012;**140**(5):541-555

[40] Lishko PV, Procko E, Jin X, Phelps CB, Gaudet R. The Ankyrin Repeats of TRPV1 Bind Multiple Ligands and Modulate Channel Sensitivity. Neuron. 2007;**54**(6):905-918

[41] Numazaki M, Tominaga T, Takeuchi K, Murayama N, Toyooka H, Tominaga M. Structural determinant of TRPV1 desensitization interacts with calmodulin. Proceedings of the National Academy of Sciences of the United States of America. 2003;**100**(13):8002-8006

[42] Moqrich A, Hwang SW, Earley TJ, Petrus MJ, Murray AN, Spencer KSR, et al. Impaired thermosensation in mice lacking TRPV3, a heat and camphor sensor in the skin. Science. 2005;**307**(5714):1468-1472

[43] Xiao R, Tang J, Wang C, Colton CK, Tian J, Zhu MX. Calcium Plays a Central Role in the Sensitization of TRPV3 Channel to Repetitive Stimulations. Journal of Biological Chemistry. 2008;**283**(10):6162-6174

[44] Shi D-J, Ye S, Cao X, Zhang R, Wang K. Crystal structure of the N-terminal ankyrin repeat domain of TRPV3 reveals unique conformation of finger 3 loop critical for channel function. Protein & Cell. 2013;**4**(12):942-950

[45] Vyklicka L, Boukalova S, Macikova L, Chvojka S, Vlachova V. The human transient receptor potential vanilloid 3 channel is sensitized via the ERK pathway. Journal of Biological Chemistry. 2017;**292**(51):21083-21091

[46] Latorre R, Brauchi S, Orta G, Zaelzer C, Vargas G. ThermoTRP channels as modular proteins with

allosteric gating. Cell Calcium. 2007;**42**(4-5):427-438

[47] Yao X, Kwan H-Y, Huang Y. Regulation of TRP Channels by Phosphorylation. Neuro-Signals. 2005;**14**(6):273-280

[48] Huang J, Zhang X, McNaughton PA. Modulation of temperature-sensitive TRP channels. Seminars in Cell & Developmental Biology. 2006;**17**(6):638-645

[49] Liu B, Yao J, Zhu MX, Qin F. Hysteresis of gating underlines sensitization of TRPV3 channels. Journal of General Physiology. 2011;**138**(5):509-520

[50] Lee SE, Lee SH. Skin Barrier and Calcium. Annals of Dermatology. 2018;**30**(3):265-275

[51] Baroni A, Buommino E, De Gregorio V, Ruocco E, Ruocco V, Wolf R. Structure and function of the epidermis related to barrier properties. Clinics in Dermatology. 2012;**30**(3):257-262

[52] Cheng X, Jin J, Hu L, Shen D, Dong X, Samie MA, et al. TRP Channel regulates EGFR signaling in Hair Morphogenesis and Skin Barrier Formation. Cell. 2010;**141**(2): 331-343

[53] Wang Y, Li H, Xue C, Chen H, Xue Y, Zhao F, et al. TRPV3 enhances skin keratinocyte proliferation through EGFR-dependent signaling pathways. Cell Biology and Toxicology. 2020;**37**(2):313-330. DOI: 10.1007/s10565-020-09536-2

[54] Bourinet E, Altier C, Hildebrand ME, Trang T, Salter MW, Zamponi GW. Calcium-permeable ion channels in pain signaling. Physiological Reviews. 2014;**94**(1):81-140

[55] Caterina MJ, Julius D. The vanilloid receptor: a molecular gateway to the pain pathway. Annual Review of Neuroscience. 2001;**24**:487-517

[56] Binshtok AM, Bean BP, Woolf CJ. Inhibition of nociceptors by TRPV1-mediated entry of impermeant sodium channel blockers. Nature. 2007;**449**(7162):607-610

[57] Caterina MJ, Leffler A, Malmberg AB, Martin WJ, Trafton J, Petersen-Zeitz KR, et al. Impaired nociception and pain sensation in mice lacking the capsaicin receptor. Science. 2000;**288**(5464):306-313

[58] Vandewauw I, De Clercq K, Mulier M, Held K, Pinto S, Van Ranst N, et al. A TRP channel trio mediates acute noxious heat sensing. Nature. 2018;**555**(7698):662-666

[59] Jordt SE, Tominaga M, Julius D. Acid potentiation of the capsaicin receptor determined by a key extracellular site. Proceedings of the National Academy of Sciences of the United States of America. 2000;**97**(14):8134-8139

[60] Gohin S, Decorps J, Sigaudo-Roussel D, Fromy B. Sensory nerves contribute to cutaneous vasodilator response to cathodal stimulation in healthy rats. Microvascular Research. 2015;**101**:103-110

[61] Marics I, Malapert P, Reynders A, Gaillard S, Moqrich A. Acute Heat-Evoked Temperature Sensation Is Impaired but Not Abolished in Mice Lacking TRPV1 and TRPV3 Channels. PLoS One. 2014;**9**(6):e99828

[62] Mandadi S, Sokabe T, Shibasaki K, Katanosaka K, Mizuno A, Moqrich A, et al. TRPV3 in keratinocytes transmits temperature information to sensory neurons via ATP. Pflügers Archiv. 2009;**458**(6):1093-1102

[63] Huang SM, Lee H, Chung M-K, Park U, Yu YY, Bradshaw HB, et al. Overexpressed Transient Receptor Potential Vanilloid 3 Ion Channels in Skin Keratinocytes Modulate Pain Sensitivity via Prostaglandin E2. The Journal of Neuroscience. 2008;**28**(51):13727-13737

[64] Fromy B, Josset-Lamaugarny A, Aimond G, Pagnon-Minot A, Marics I, Tattersall GJ, et al. Disruption of TRPV3 Impairs Heat-Evoked Vasodilation and Thermoregulation: A Critical Role of CGRP. Journal of Investigative Dermatology. 2018;**138**(3):688-696

[65] Michalek IM, Loring B, John SM. A systematic review of worldwide epidemiology of psoriasis. Journal of the European Academy of Dermatology and Venereology. 2017;**31**(2):205-212

[66] Lowes MA, Suárez-Fariñas M, Krueger JG. Immunology of Psoriasis. Annual Review of Immunology. 2014;**32**:227-255

[67] Rendon A, Schäkel K. Psoriasis Pathogenesis and Treatment. International Journal of Molecular Sciences. 2019;**20**(6):1475

[68] Nattkemper LA, Tey HL, Valdes-Rodriguez R, Lee H, Mollanazar NK, Albornoz C, et al. The Genetics of Chronic Itch: Gene Expression in the Skin of Patients with Atopic Dermatitis and Psoriasis with Severe Itch. The Journal of Investigative Dermatology. 2018;**138**(6):1311-1317

[69] Zhou Y, Follansbee T, Wu X, Han D, Yu S, Domocos DT, et al. TRPV1 mediates inflammation and hyperplasia in imiquimod (IMQ)-induced psoriasiform dermatitis (PsD) in mice. Journal of Dermatological Science. 2018;**92**(3):264-271

[70] Fits L v d, Mourits S, JSA V, Kant M, Boon L, Laman JD, et al. Imiquimod-Induced Psoriasis-Like Skin Inflammation in Mice Is Mediated via the IL-23/IL-17 Axis. The Journal of Immunology. 2009;**182**(9):5836-5845

[71] Nociceptive Sensory Neurons Drive Interleukin-23 Mediated Psoriasiform Skin Inflammation. Nature. 2022;**510**(7503):157-161

[72] Roblin D, Yosipovitch G, Boyce B, Robinson J, Sandy J, Mainero V, et al. Topical TrkA Kinase Inhibitor CT327 is an Effective, Novel Therapy for the Treatment of Pruritus due to Psoriasis: Results from Experimental Studies, and Efficacy and Safety of CT327 in a Phase 2b Clinical Trial in Patients with Psoriasis. Acta Dermato-Venereologica. 2015;**95**(5):542-548

[73] Zhang X, Huang J, McNaughton PA. NGF rapidly increases membrane expression of TRPV1 heat-gated ion channels. The EMBO Journal. 2005;**24**(24):4211-4223

[74] Raychaudhuri SP, Raychaudhuri SK. Role of NGF and neurogenic inflammation in the pathogenesis of psoriasis. Progress in Brain Research. 2004;**146**:433-437

[75] Seo S, Kim S, Kim S-E, Chung S, Lee S. Enhanced Thermal Sensitivity of TRPV3 in Keratinocytes Underlies Heat-Induced Pruritogen Release and Pruritus in Atopic Dermatitis. Journal of Investigative Dermatology. 2020;**140**(11):2199-2209.e6

[76] Szöllősi AG, Vasas N, Angyal Á, Kistamás K, Nánási PP, Mihály J, et al. Activation of TRPV3 Regulates Inflammatory Actions of Human Epidermal Keratinocytes. Journal of Investigative Dermatology. 2018;**138**(2):365-374

[77] Zhao J, Munanairi A, Liu X-Y, Zhang J, Hu L, Hu M, et al. PAR2

Mediates Itch via TRPV3 Signaling in Keratinocytes. The Journal of Investigative Dermatology. 2020;**140**(8):1524-1532

[78] El-Ghareeb MI, Helmy A, Al Kazzaz S, Samir H. Serum TSLP is a potential biomarker of psoriasis vulgaris activity. Psoriasis (Auckl). 2019;**9**:59-63

[79] Yoshioka T, Imura K, Asakawa M, Suzuki M, Oshima I, Hirasawa T, et al. Impact of the Gly573Ser Substitution in TRPV3 on the Development of Allergic and Pruritic Dermatitis in Mice. Journal of Investigative Dermatology. 2009;**129**(3):714-722

[80] Imura K, Yoshioka T, Hirasawa T, Sakata T. Role of TRPV3 in immune response to development of dermatitis. Journal of Inflammation (Lond). 2009;**6**:17

[81] Bang S, Yoo S, Yang T, Cho H, Hwang S. 17(R)-resolvin D1 specifically inhibits transient receptor potential ion channel vanilloid 3 leading to peripheral antinociception. British Journal of Pharmacology. 2012;**165**(3):683-692

[82] Kemény Á, Kodji X, Horváth S, Komlódi R, Szőke É, Sándor Z, et al. TRPA1 Acts in a Protective Manner in Imiquimod-Induced Psoriasiform Dermatitis in Mice. The Journal of Investigative Dermatology. 2018; **138**(8):1774-1784

[83] Greco C, Leclerc-Mercier S, Chaumon S, Doz F, Hadj-Rabia S, Molina T, et al. Use of Epidermal Growth Factor Receptor Inhibitor Erlotinib to Treat Palmoplantar Keratoderma in Patients With Olmsted Syndrome Caused by TRPV3 Mutations. JAMA Dermatology. 2020;**156**(2):191-195

[84] Smith GD, Gunthorpe MJ, Kelsell RE, Hayes PD, Reilly P, Facer P, et al. TRPV3 is a temperature-sensitive vanilloid receptor-like protein. Nature. 2002;**418**(6894):186-190

[85] Cheng W, Yang F, Liu S, Colton CK, Wang C, Cui Y, et al. Heteromeric heat-sensitive transient receptor potential channels exhibit distinct temperature and chemical response. The Journal of Biological Chemistry. 2012;**287**(10):7279-7288

[86] Price E, Gianfrancesco O, Harrison PT, Frank B, Bubb VJ, Quinn JP. CRISPR Deletion of a SVA Retrotransposon Demonstrates Function as a cis-Regulatory Element at the TRPV1/TRPV3 Intergenic Region. IJMS. 2021;22(4):1911.

Chapter 4

TRPV Family Ion Channels in the Mammary Epithelium: Role in Normal Tissue Homeostasis and along Breast Cancer Progression

Sari Susanna Tojkander

Abstract

Calcium homeostasis directs various intracellular cascades and therefore strict spatio-temporal control of calcium influx is also crucial for diverse physiological processes. In the mammary gland, calcium is important for the specialized tasks of this organ during lactation, but it also guides other structural and functional features of the mammary epithelium and in this way the maintenance of the whole tissue. Transient receptor potential, TRP, family ion channels are cationic channels, permeable to both monovalent and divalent cations and play a role in the influx of calcium mainly through the plasma membrane. These channels also represent vital calcium entry routes in the mammary epithelium and may thus act as central players in the preservation of calcium balance within this tissue. Moreover, TRP family channel proteins are abnormally expressed in breast cancers and may promote cancer progression through deregulation of intracellular signaling, consequently triggering several hallmarks of cancer. This chapter concentrates on the role of transient receptor potential vanilloid, TRPV, a subfamily of proteins in the calcium-dependent functions of normal mammary epithelium and the evident role of these channel-forming proteins along breast cancer progression.

Keywords: TRP, TRPV, calcium, calcium signaling, mammary epithelium, epithelial integrity, breast cancer, invasion

1. Introduction

In adult individuals, the mammary gland is composed of bilayered epithelial structures, forming a branched ductal tree within an adipocyte-rich stroma [1]. These tree-like structures consist of distinct epithelial cell populations that form secretory alveoli, organized into lobules and a branched network of ductal structures. Development of the mammary epithelial cell populations within these structures, occurs hierarchically through specific intermediates and coordinated expression of several lineage-specific markers [2–4]. Functionally these distinct cell populations are organized into an inner luminal epithelial (LE) cell layer and outer basal cell layer, the basal layer containing both mature myoepithelial (ME) and stem/progenitor cell populations [5]. These specific cell populations within the bilayered mammary epithelium can be distinguished by the expression of various markers, including the cytokeratin expression pattern [6].

The basal cell layer is responsible for the regenerative potential of the mammary epithelium due to the colonization of the mammary stem cells with multilineage potential, within this compartment [5, 7, 8]. Contractile ME cells, localized to the same cell layer, provide a niche for these stem cells. Additionally, ME cells have an important role in synthesizing and maintaining normal basement membrane (BM), controlling polarization and proliferation of the LE cells as well as directing branching and differentiation of the developing structures [9, 10]. Upon gestation, epithelial cell populations further undergo directed differentiation and proliferation, consequently leading to side-branching and formation of alveolar, lactating units within lobular clusters [11–13]. In such functionally mature mammary epithelial structures, the inner luminal cell population produces and secretes milk into the lumen [14, 15], whereas the outer, smooth muscle actin (α-SMA)-expressing myoepithelial cells provide contractile forces for milk ejection in response to oxytocin [14, 16, 17]. When lactation is over, the alveolar cells undergo programmed cell death and the epithelium is returned to its pregestational state [18–20].

Calcium is crucial for various physiological processes through activation of specific intracellular cascades and by modulating the integrity of cellular junctions [21, 22]. Alterations in the activity or expression levels of different Ca^{2+} channels, or factors involved in their regulation can therefore significantly change cellular responses to various cues that direct tissue homeostasis [23, 24]. Consequently, deregulation of calcium signaling is therefore associated with several pathological conditions, including cancers. In cancers, abnormal calcium signaling has been linked to high proliferation, inhibition of apoptosis and invasive migration through the epithelial-to-mesenchymal transition, EMT [25, 26]. As for any other tissue, calcium signaling is likewise crucial for the regulation of mammary epithelium, its various calcium-dependent intracellular functions, the integrity of the epithelial sheets and mammary tissue-specific task, lactation [27]. The functional maintenance of the bilayered mammary epithelium is importantly also guided by various hormones and growth factors, which may also cooperate with calcium-triggered pathways [28–30]. In this review, the role of TRPV, vanilloid subgroup of transient receptor potential family ion channels are discussed in respect of their significance in the regulation of normal mammary epithelial homeostasis and along breast cancer progression.

2. TRPV family channels

TRPV channel proteins belong to the transient receptor potential, TRP, the family of proteins [23]. This superfamily of proteins is formed by over 30 different cationic channel proteins, which are further divided into seven subfamilies: TRPV (vanilloid), TRPA (ankyrin), TRPC (canonical), TRPM (melastatin), TRPML (mucolipin), TRPN (NOMPC), and TRPP (polycystin) families [23, 31]. TRP proteins possess crucial functions in various tissues, in both non-excitable cells as well as in the cells of the nervous system [23, 31]. The members of this family display both structural and functional similarities and many of them are voltage- and temperature-sensitive for functioning as sensors in the peripheral and central nervous systems. Besides these, they can sense various other extracellular cues, both biochemical and physical ones, leading to their spatio-temporal activation, ion influx and adjustment of the downstream signaling cascades [32, 33].

The subfamily of TRPVs has six members, TRPV1-6 that can form both homo-and heterodimeric channels. Of these subfamilies, TRPV1-4 do not display high Ca^{2+}-selectivity [34–37]. On the opposite, TRPV5 and TRPV6 are highly selective for Ca^{2+} [38–40]. Most of these TRPV family members can sense and respond to various stimuli, consequently activating multiple intracellular signaling cascades [31, 41, 42]. TRPV1 is found in the plasma membrane and prominently expressed in sensory neurons but it is also clearly expressed in other cell types [43]. TRPV1 is involved in nociception and triggered by heat, pH and some compounds, including capsaicin [44–46]. TRPV2 is found in all tissues, and is highly expressed in sensory neurons. Its main localization in cells is not at the plasma membrane but in the intracellular membranes [47–49]. TRPV2 displays various physiological functions through its actions as a thermo-, lipid- and mechanosensor. Additionally, it can also respond to growth factors, hormones and cytokines [50–52], leading to a wide range of functions that play a role in healthy tissues and in pathophysiological conditions.

Both TRPV3 and TRPV4 are highly ubiquitous and they are noticeably expressed in epithelial tissues [35, 37, 53]. TRPV3 is a non-selective cation channel and is especially abundant in the skin keratinocytes [54]. It can sense temperature and plays a role in various tasks, including maintenance of the skin barrier function, wound healing, pain sensation and itch [53, 55]. Therefore, TRPV3 seems to be particularly important for the skin health. Like TRPV3, TRPV4 is an abundant cationic channel in the epithelial tissues and can trigger ion-influx upon various cues, such as mechanical stretching, osmolarity and heat [56–60]. The activity of TRPV4 has been associated with various physiological functions, it has an important role in cell volume regulation, homeostasis of the vasculature, central nervous system and as a mechanosensor in a wide array of tissues [37, 61, 62].

Unlike the other TRPVs, TRPV5 and its close relative TRPV6 are the only highly calcium-selective channels among TRPVs and the whole TRP superfamily [38–40]. TRPV5 is highly expressed in the kidney, while TRPV6 has a broad expression pattern in some different tissues. TRPV5 and TRPV6 constitute the apical Ca^{2+} entry mechanism for active calcium transport in the kidney and intestine, respectively. Their roles in the active Ca^{2+}-reabsorption and maintenance of cellular Ca^{2+}-homeostasis are essential, loss of these proteins leading to reduced bone thickness, defects in the intestinal calcium absorption, reduced fertility, and hypocalcemia [63–67]. Interestingly, TRPV5 and TRPV6 are under the regulation of 1,25-dihydroxyvitamin D3, and hormones, such as parathyroid hormone, estrogen, and testosterone may participate in fine-tuning the calcium-uptake [68–73].

The hormonal regulation of TRPV channels has mainly concentrated on the role of sex hormones, which can impact the expression of ion channels either directly or indirectly through intracellular signaling [74, 75]. Progesterone, a steroid hormone, is known to elevate TRPV6 levels in mammary carcinoma cells [76]. In human mammary epithelium, progesterone receptor, PR, is expressed in both luminal and basal epithelial cell populations, and it promotes the proliferation of the basal mammary epithelial cells. Luminal PR may also promote the proliferation of neighboring cells through paracrine signaling mechanisms [77]. In addition to TRPV6, TRPV4 is under the control of progesterone receptors in the mammary gland, airways and smooth muscle cells of the vasculature [78]. In the case of TRPV4, progesterone was found to decrease both mRNA and protein levels of TRPV4, while silencing of PR led to increased level and activity of TRPV4 in the T47D mammary epithelial cell model [78].

In adult individuals, the PR-positive cells are usually also ERα positive [79] and estrogen acts through ERα to induce the expression of PR [80, 81]. This interconnection between the hormone receptors and specific TRPV channel proteins should be further assessed in future studies, as they may also play a role in the disease progression.

3. TRPV channels in the structural maintenance of the mammary epithelium

Calcium signaling is known to direct developmental processes and is also crucial for both structural and functional maintenance of the mammary epithelium [82]. Different TRP family proteins serve as important calcium influx routes in the mammary epithelium and may thus act as central players in the maintenance of the mammary epithelium through calcium homeostasis.

Among the TRPV family channel members, the TRPV4 channel is probably the most well studied in respect of its role in epithelial integrity through the regulation of adherens- and tight junction proteins [83–90]. In a mouse mammary epithelial cell line, HC11, TRPV4 localizes at the basolateral membrane to regulate calcium influx and permeability [86]. This TRPV4-mediated Ca^{2+}-intake is known to trigger activation of some calcium-dependent voltage-gated potassium channels, BK channels, that have a major role in tight junction regulation through at least claudin family proteins. Mechanistically, TRPV4-mediated calcium influx leads to two separate cellular events: A fast elevation in the transcellular conductance via the activation of apically-located large BK potassium channels and a slower increase in paracellular permeability for small soluble molecules. Associated with these alterations, down-regulation of several claudin family tight junction proteins was detected together with large break formation in the tight junction strands [86]. In contrast, studies by Islam et al. showed that TRPV4 can also positively affect the expression of tight junction proteins through X-box-binding protein 1, XBP1, in the mammary epithelial cells upon heat induction [89]. Besides TRPV4, also TRPV6 may play a role in the homeostasis of the mammary epithelium, both during differentiation and maintenance of the intact epithelial structures: Zinc finger homeobox 3 (ZFHX3) is a transcription factor that directs numerous cellular processes, including differentiation. ZFHX3 was found to regulate calcium homeostasis in the mammary epithelium through positive regulation of TRPV6, leading to differentiation of MCF10A mammary epithelial cells in the 3D environment [91]. These observations support the role of TRPV6-mediated calcium influx in the differentiation and maintenance of the mammary epithelium, downstream of ZFHX3. As ZFHX3 is also linked to the function of hormones, including progesterone which can upregulate TRPV6, it would be interesting to investigate the possible connection between them. Furthermore, TRPV6 seems to be important for the maintenance of the junctional integrity of the mammary epithelium [92]. TRPV6 was found to localize at the cell-cell junctions together with adherens junction protein E-cadherin and its depletion led to the loss of epithelial integrity as detected with both MCF10A and 184A1 mammary epithelial cell cultures, treated with TRPV6 siRNA. This could be at least partially through the regulation of peripheral actomyosin bundles that maintain junctional tension as TRPV6 depletion affected pathways upstream of actomyosin assembly [92]. While there is evidence for the role of TRPV4 and TRPV6 in the structural maintenance of the mammary epithelium, the possible role of the other TRPV channel family members have not been properly assessed in this respect, at least in the mammary epithelial model. Additionally, it may be that these channel proteins respond differently to distinct cues to regulate the junctional integrity in the epithelial sheets. At least in the case of TRPV4, there seems to be dual modulation depending on the initial cues.

4. Functions of TRPV channels along gestation and lactation

In the course of gestation, the mammary gland and its epithelial structures undergo major architectural changes, leading to the formation of milk-producing alveolar structures. These morphological events are jointly guided by hormones and growth factors, alterations in the physical microenvironment as well as the paracrine signaling in between the mammary stroma and the bilayered epithelium [93–95]. Coinciding with the formation of alveoli, TRPV4 mRNA levels are known to be increased at the day 15 of gestation and to be downregulated immediately after lactation [89]. These findings suggest that at least TRPV4 could have a role in the pregnancy-linked developmental processes within the mammary gland. While other, TRPV channels are also responsive to hormones and changes in the mechanical microenvironment, their role along gestation-linked epithelial changes have not been assessed.

During lactation, the maternal calcium and magnesium homeostasis encounter significant alterations due to excessive need of the divalent cation Ca^{2+} in breast milk. Consequently, demineralization of the skeleton is observed together with changes in both renal and intestinal Ca^{2+} transport [96]. For this, several proteins, playing a role in the transcellular Ca^{2+} and Mg^{2+} transportation are upregulated along lactation. Vitamin D also contributes to this process by inducing intestinal hyperabsorption [97]. TRPV5 is highly expressed in the kidney epithelium, in the distal convoluted tubules and connecting tubules [98]. Structurally similar TRPV6 is more widely expressed but exhibits prominent expression in the intestine epithelium [99, 100]. Moreover, both TRPV5 and 6 are Ca^{2+}-selective and also vitamin D-responsive [101], and in line with this connection to lactation-induced alterations in Ca^{2+}-homeostasis, they are also upregulated in renal and intestinal epithelium upon lactation [97]. Furthermore, prolactin is known to regulate both vitamin D metabolism and induce TRPV6 levels to regulate calcium intake during lactation [102]. TRPV5 and TRPV6 thus participate to lactation by enabling the excessive need of calcium during this physiological phase.

Production of milk is triggered by heat as mammary epithelial cells can activate their milk generation at 39 degrees [103, 104]. Mammary epithelial cells also undergo heat-evoked proliferation and differentiation [105]. Interestingly, many TRP channels act in sensing heat and from the vanilloid subfamily of proteins, TRPV1-4 acts as major thermosensors [45, 106–108]. Upon heat-treatment, TRPV4 is also able to activate the expression of milk protein beta-casein and tight junction (TJ)-associated proteins, Zonula occludens-1 (ZO-1), Claudin 3 (Cldn3) and Occludin (Ocln) [89]. Permeability of TJs is known to be modulated upon milk production and immediately after parturition [109], and this feature may thus be dependent on TRPV4. Heat stress is also known to induce unfolded protein response, UPR [110] and UPR-associated transcription factor XBP1 plays a role in the differentiation of mammary epithelium together with the expression of milk protein beta-casein [104, 111]. Intriguingly, recent work by Islam et al. proposes that TRPV4 acts through XBP1 [89]. Besides heat, TRPV4 is activated by mechanical changes and stretching in the cell environment that are also known to take place along lactation. In addition to TRPV4, the TRPV2 channel can play a role in lactation as it localizes to oxytocinergic neurons [112].

After lactation is over, the milk-producing structures regress to the pre-pregnancy state in a complicated reverse action, involution [113]. Ca^{2+}-dependent signaling may also impact this transfer from lactation to involution [14]. Whether any of the TRPV family members play a role in this process, remains to be studied.

5. Abnormal expression of TRPV channels along breast cancer progression

The characterization of breast cancers is based on different criteria, including the histopathological evaluation, grading and staging as well as defining the expression of estrogen (ER), progesterone (PR), and epidermal growth factor (HER2) receptors [114]. Additionally, gene expression profiles can be used to determine the molecular subtypes, which can be Basal-like, HER2-enriched, Claudin-low, Luminal A, Luminal B, or Normal-like. The heterogeneity of breast cancer as a disease is well seen also on the differences in Ca^{2+}-channel expression that vary greatly in between specific breast cancer subtypes. Often the levels or activity of plasma membrane-embedded calcium channels can also reflect the metastatic potential and prognosis of distinct mammary carcinomas [26]. Abnormal activity of the Ca^{2+}-channels in breast cancers could potentially take place due to mutations, deregulation of the channel gating or changes in the expression levels, triggering Ca^{2+}-influx in unfavorable patterns, both spatially and temporally. As several calcium channels can respond to a wide variety of biochemical and mechanical cues in their microenvironment, any alterations in such could lead to deregulated calcium channel activity to sustain an elevated or abnormally low calcium entry. Additionally, a variety of the plasma membrane-associated calcium channels could be deregulated at the same time, within similar cancer types, further cooperating in adverse processes along the course of neoplastic progression.

TRPV channel family, among the other TRP family members, has been linked to the progression of a variety of human cancers [115, 116]. These cationic channels can also mediate Ca^{2+}-influx and have been shown to contribute to several hallmarks of cancers, including the potential to proliferate, resistance to apoptosis, angiogenesis, and invasion [117, 118]. Additionally, these channel proteins may have different roles, as either cancer promoters or suppressors, depending on the cancer type and its genetic background as well as the expression levels of distinct channel proteins. The primary Ca^{2+}-triggered pathways that play a role in promoting these cancer-associated features through specific TRP channels, include CaMKII, NF-κB, calpains and calcineurin pathways [119, 120], but other less studied signaling cascades may as well be involved.

While the members of TRPV family channels are frequently deregulated in many cancers and associated with certain cancer-specific cellular features, their regulation along the breast cancer progression is still poorly understood. TRPV1 channel is often upregulated in breast cancers and its high expression correlates with the tumor grade [121]. Some studies have shown no differences in between distinct breast cancer sub-types and expression levels of TRPV1 [122–124]. Aggregated TRPV1 in the intracellular compartment has, however, been linked to poor prognosis in breast cancer patients [125]. TRPV2 expression also seems to display oncogenic activity in various cancers [126, 127]. In triple-negative breast cancers (TNBCs), TRPV2 levels are especially prominent but correlate interestingly with high relapse-free survival in this case [122, 128]. Additionally, the study by Elbaz et al. [128], proposed the therapeutic potential of high TRPV2 to elevate the uptake and efficacy of chemotherapeutic agents in patients with TNBC. The role of TRPV4 in cancer progression has been investigated by several labs and its expression in breast cancers is highest in the basal-like cancer subtype [122, 129]. High TRPV4 expression has also been detected in IHC stainings from the metastatic lesions of invasive ductal carcinomas and its levels correlate with the tumor grade and size [130].

TRPV6 channel is likewise overexpressed in various cancers, including cancers of the mammary tissue [76, 131–134]. The levels of overexpressed TRPV6 vary a lot depending on the breast cancer subtype, and as with the TRPV4 channel also TRPV6 levels are highest in the basal-like breast cancers and HER2-enriched molecular subtypes [76, 135, 136]. In line with this, ER receptor-negative breast cancers and cancer cell lines with several overlapping features with the basal and HER2-enriched subtypes display significant amounts of TRPV6 [136]. High TRPV6 in the patients is also associated with lower survival in comparison to patients that express lower TRPV6 levels [136].

Of the TRPV family, TRPV3 and TRPV5 have been studied to less extent. TRPV3 is known to be expressed at low levels in different types of breast cancer subtypes and its possible association with cancer progression has not been well assessed [122]. Likewise, there are no reports on TRPV5 and its link to the progression of distinct breast cancer types [122]. While these two subtypes may not be important in respect of breast cancer progression, more studies are needed on the field to understand how the deregulation of the other TRPV forms takes place along cancer progression and whether for instance hormonal regulation or stromal changes could impact their expression and activity.

6. TRPV family channels: implications for cancer cell-associated features along breast cancer progression

6.1 Excessive proliferation

Various studies have shown the significance of calcium signaling in the uncontrolled proliferation of cancer cells [25, 137, 138]. Various plasma membrane-embedded calcium channels are acting as major sources of calcium for the regulation of such pathways that lead to elevated cell amounts [139, 140]. Among the TRP family, also TRPV channels contribute to these processes, related to malignant growth [120].

TRPV1 channel can mediate both Ca^{2+} and Na^{+}-influx and trigger cell proliferation by two separate mechanisms: It can contribute to the activation of serine-threonine kinase Akt as well as to the activation of ERK1/2 downstream of the epidermal growth factor (EGFR) [141]. However, studies in the MCF-7 breast cancer cell line show that both agonists and antagonists of the TRPV1 channel can inhibit cell growth through yet unidentified mechanisms [142]. Thus, it may be that the balance in the expression of this protein is important for controlled cell proliferation through distinct intracellular pathways in a cell type-specific manner. MCF-7 cell line has also been utilized as a model to study TRPV2 in respect of cell proliferation. TRPV2 was shown to be responsive to insulin-like growth factor-I (IGF-I) [143] and tranilast, an anti-inflammatory agent, was reported to inhibit IGF-1-induced cell growth by blocking the calcium influx through TRPV2 [144]. Significant overexpression of TRPV4 is also linked to breast cancers and this seems to correlate with the tumor grade and size, leading to poor survival [122, 130, 145, 146]. However, evidence from studies performed with 4 T07, MDA-MB-231 and MDA-MB-468 breast cancer cell lines show that TRPV4 is dispensable for the proliferative potential of these specific breast cancer cell lines, since its silencing or pharmacological inhibition was not anti-proliferative [145, 147]. In contrast, high expression of TRPV6 is linked to the proliferation through Ca^{2+}-triggered intracellular pathways and the high levels also act as prognostic factors together with potential resistance to chemotherapy [76, 134, 135]. Depletion of TRPV6 from the T47D breast cancer cell line, displaying high endogenous TRPV6, also decreases the proliferation of these cells [76, 136]. The precise mechanisms behind this are not understood but may involve PI3K/pAKT pathway that regulates cell proliferation, survival and therapeutic resistance in some breast cancer subtypes, including the HER2-enriched subtype. In line with this, depletion of TRPV6 was associated with lower levels of active, phosphorylated AKT in HCC-1569 breast cancer cells [148]. Studies in breast cancer cell lines, MCF-7 and MDA-MB-231, additionally showed the link in between TRPV6 and PI3K/Akt pathway as a functionally auto-inhibitory intramolecular interaction between S5 and S6 helices of TRPV6 was shown to contribute to TRPV6/PI3K association and the activation of PI3K/Akt/GSK-3β pathway [149].

TRPV6 activity and expression are known to be controlled by estrogen, progesterone, and 1,25-vitamin D that all play a role in the proliferation of breast cancer cells [76, 120]. Treatment of breast cancer cell line T47D with estrogen receptor antagonist, tamoxifen, also led to lower activity and expression of TRPV6 calcium channel protein. Further, the effect of tamoxifen on the functionality of TRPV6 was shown in EYFP-C1-TRPV6-transfected MCF7 breast cancer cells by Fura-2 calcium imaging [150]. Calcium levels in the transfected cells were found to be higher than in non-transfected cells and the calcium levels were lowered by 50% with tamoxifen-treatment. Interestingly, tamoxifen also played a role in TRPV6 inhibition in MDA-MB-231 cells that are estrogen receptor-negative [150], suggesting a direct impact on TRPV6-mediated Ca^{2+}-influx. Besides tamoxifen, TRPV6 activity can also be negatively regulated by a protein called Numb1 [151]. Numb1 is maybe more known for its role in the stabilization of tumor suppressor protein p53 [152], affecting both cell cycle progression and apoptosis. Studies on the Numb1-TRPV6 link in MCF-7 breast cancer cells showed that Numb1-depleted cells displayed elevated TRPV6 expression and calcium influx as well as enhanced proliferation. TPV6 thus has interesting connections to the pathways of the major tumor suppressor protein as well as to hormones that play a role in breast cancer progression through the proliferative potential of the cells.

6.2 Resistance to apoptosis

Apoptosis can be characterized as a programmed cell death process, which leads to the fragmentation of DNA. This strictly controlled process can take place through cell death-receptors or through mitochondria-mediated apoptotic pathways [153]. Apoptosis is also controlled by calcium-dependent pathways [154–156]. Changes in intracellular Ca^{2+}-levels are known to influence the two major apoptotic pathways through gene expression [157–161]. For instance, the calcium/calmodulin-dependent signaling cascades can affect the balance in between cell cycle progression and apoptosis [160].

TRPV1-triggered calcium influx has been shown to act as a determinator of the balance in between cell proliferation and apoptosis. TRPV1-mediated apoptosis can take place through the mitochondrial mechanism, while its proliferation-supportive actions usually involve other cell membrane receptors or specific intracellular signaling cascades [141]. Studies with MCF-7 breast cancer cell line have also shown that high TRPV1 sensitizes cells to programmed cell death, induced by TRPV1 activator capsaicin [162, 163]. Likewise, capsaicin is involved in the induction of cell death in breast cancer cell line, SUM149PT through TRPV1 activation [121].

The role of TRPV4 in apoptosis has as well been investigated during the past few years and these studies support the role of high TRPV4 expression in inducing cell death. In breast cancer cell lines, MDA-MB-468 and HCC1569, activation of TRPV4 by pharmacological compounds reduced the viability of the cells [147]. Both cell lines display high endogenous TRPV4 levels and its activation was able to promote cell death by apoptosis or oncosis, while the same phenomenon was not detected in breast cancer cell lines with low TRPV4 levels. Moreover, the studies by Peters et al. found that TRPV4 activation has therapeutical relevance in vivo and can inhibit the growth of tumors [147]. Similarly, to TRPV4, overexpression of TRPV2 and its pharmacological activation with cannabidiol have been linked to inducing cytotoxic impact in SUM159 and MDA-MB-231 breast cancer cells via doxorubicin-treatment [128]. In contrast, the TRPV6 calcium channel seems to act oppositely and its high levels are rather protecting from apoptosis: TRPV6 calcium channel is known to get transported to the plasma membrane in an Orai1-mediated mechanism to control the survival of the cancer cells [164]. On the other hand, TRPV6 depletion from breast cancer cells with high expression of this protein can be used for decreasing the viability of the cells, as shown by studies in T47D breast cancer cells [76].

6.3 Tumor microenvironment and angiogenesis: connection to TRPV channels

The tissue microenvironment undergoes drastic alterations along breast cancer progression [165–167]. Besides stiffness and composition of the matrix, there are also changes for instance in the amount of growth factors and acidicity of the environment that may trigger specific calcium channels [168, 169]. How TRPV channels, among other ion channels on the plasma membrane, respond to such cancer-linked cues from the extracellular space, is poorly understood. Additionally, stromal cells, such as fibroblasts, immune cells, or adipocytes that also express channel proteins, may be functionally altered and contribute to abnormal signaling from the stroma.

At least TRPV4 and TRPV6 are known to be responsive to stromal stiffening [92, 170–173] and could be triggered by cancer-associated mechanical changes in the extracellular space. Furthermore, TRPV4 has been shown to control the expression of some extracellular matrix proteins, in this way contributing itself to the stiffness of the environment [130]. Stiffening may impact various processes along cancer progression and one of these features is the growth of new vasculature, angiogenesis. The first evidence that TRPV4 could also be involved in angiogenesis along breast cancer progression was presented in the work by Fiorio Pla et al. [174]. The authors discovered the role of TRPV4 in mediating arachidonic acid (AA)-promoted migration of endothelial cells (ECs), derived from breast tumors. These endothelial cells displayed high endogenous TRPV4 and were enhancing the migration of ECs, a key step in the growth of new vessels. This step could be inhibited by antagonist or siRNAs against TRPV4 and the opposite was detected with TRPV4 stimulation [174].

Support for the role of TRPV4 in angiogenesis has also been shown in studies by Adapala et al. [170]. TRPV4 seems to control the mechanosensitivity of tumor endothelial cells (TECs), and the angiogenetic process all the way to the maturation of the vessels. Interestingly, the authors found that these TECs display lower TRPV4 levels than normal endothelial cells, leading to angiogenesis through the altered ability of the cells to sense the mechanical environment. Besides, they discovered that normalizing TRPV4 levels could be acting as an anti-angiogenetic therapy to normalize the vasculature and enhance drug efficiency. Moreover, studies by Thoppil et al. have shed light on the mechanisms that TRPV4 could utilize in the regulation of the angiogenetic process [175]. These studies also linked low TRPV4 levels of endothelial cells to enhanced migration and disturbed angiogenesis. This could be reversed by the treatment of cells with Rho kinase inhibitor, Y-27632, suggesting that TRPV4 action in angiogenesis involves modulation of mechanosensitivity of ECs via Rho pathway [175]. Based on these data, TRPV4 may therefore be a significant regulator of angiogenesis and this information could potentially be utilized in therapeutical approaches. TRPV4 thus has an important role in the modulation of tumor stroma by affecting both its mechanical features as well as the growth of new blood vessels in the stroma. Interestingly, TRPV4 is this far the only channel protein among the TRP superfamily that has been implicated in the growth of new vessels along cancer progression.

6.4 Invasion and metastasis

Abnormal expression of distinct TRPV channels has also been linked to invasive migration and metastasis. Several TRP channel family members are connected to Rho-pathway and display the potential to promote invasive migration through Rho-dependent cytoskeletal reorganization [174, 176]. Of the TRPV family members, at least TRPV2 appears to be under the control of Rho-kinase as the treatment of breast cancer cells with Rho-inhibitors lowers the levels of TRPV2 [177]. Another factor, known to impact cell migration through activation of TRPV2, is the antimicrobial peptide, LL-37. LL-37 can influence cancer progression through various ways, including its positive impact on cancer cell migration [178]. The expression of LL-37 correlates with the expression levels of TRPV2 in breast cancer cell lines and has been shown to promote invasive migration of MDA-MB-435, MCF-7 and MDA-MB-231 cells dependently on TRPV2 [179]. Mechanistically, LL-37 increases calcium influx through TRPV2 and enhances cell migration via PI3K/AKT signaling [180]. Activation of PI3/Akt pathway as such leads to recruitment of TRPV2 into pseudopodia, impacting the migration of specific breast cancer cell types [179].

TRPV4 has also been associated with invasive migration and has been linked to EMT and lower relapse-free survival in basal breast cancers with lymph node involvement [181]. In MDA-MB-468 breast cancer cells, TRPV4-mediated calcium-influx plays an important role in the EGF-triggered EMT: activation of TRPV4 by chemical compounds was able to drive the upregulation of various EMT markers in these cells [181]. In line with these results, TRPV4 depletion from a murine mammary cancer cell line, 4T07, lowered the migration capability and 3D invasion of these normally high TRPV4-expressing cells [145]. Furthermore, determining TRPV4 levels from database information of human clinical samples as well as phosphoproteomic analyses of xenograft-derived in vitro models, indicated the role of TRPV4 in breast cancer metastasis, high expression of TPV4 in basal breast cancers and its association with poor prognosis [145]. Additionally, TRPV4 KD decreased the levels of metastatic nodules in mouse xenografts [145].

Interestingly, TRPV4 also implies to determine the stiffness of cancer cells through actin dynamics, in this way affecting deformability and metastasizing potential of breast cancer cells [130, 145]. TRPV4 was regulating the compliance of cancer cells through Ca^{2+}-mediated AKT-E-cadherin signaling [130]. Additionally, TRPV4 was involved in the expression of extracellular matrix proteins and the modeling of the matrix [130]. Knowing the mechanosensitive nature of TRPV4, there seems to be a functional feedback loop in between TRPV4 and its mechanical environment that plays a role along cancer progression. TRPV4 may therefore impact invasion and metastasis of breast cancer cells through various means.

Besides TRPV2 and -4, also TRPV6 has been linked to invasion and metastasis in breast cancers. Overexpression of TRPV6 is very common in breast carcinomas and TRPV6 levels have been shown to be very high in the invasive regions of the mammary carcinoma samples [76, 135]. The mechanisms of how TRPV6 could impact invasive progression, are not well understood. However, it seems to be linked to both actomyosin dynamics and the expression of EMT markers that could be critical along the development of invasive disease [92]. Further, inhibition of TRPV6-mediated calcium-influx by lidocaine, led to lower migration and invasion ability of the MDA-MB-231 breast cancer cells [182]. The exact molecular pathways, affecting TRPV6-mediated invasion in breast carcinomas, needs still to be further clarified.

7. Pain sensation

TRPV channels have been indicated to function in nociception, the sensation of pain [41, 44]. Although, not directly linked to the function of the mammary gland, it plays a role in breast cancer progression in the form of bone pain as a consequence of bone metastasis formation.

One of the main TRPV channels, playing a role in nociception, is TRPV1 [44, 46]. Interestingly, the formation of a tumor within a bone is known to increase the expression of TRPV1 in a specific population of dorsal root ganglion neurons [183]. In addition, TRPV1 is important for both the development and maintenance of cancer pain [184]. Likewise, it has been observed that extracellular cues within the bone microenvironment, developed during the formation of breast cancer-derived metastasis, are contributing to the pain sensation via TRPV1 activation [185]. In line with these data, experiments with rat models have revealed that when mammary carcinoma cells are injected to the rat bones, TRPV1 expression is upregulated within the dorsal root ganglion cells [184, 186]. Further, MDA-MB-231 breast cancer cells have been shown to promote sensory neuronal growth and elevate sensitivity to active TRPV1 [187]. TRPV1 may therefore be important in the sensation of pain upon metastatic breast cancer and its pharmacological targeting has also been pursued for instance by blocking the capsaicin receptor [188].

The mechanisms through which TRPV1 is induced upon bone cancer and -metastasis have been studied as well: in a rat bone cancer-pain model, utilizing mammary carcinoma cells injected to the bone cavity, TRPV1 was upregulated and activated through induction of Insulin-like Growth Factor 1, IGF-1 [184]. Additionally, TGF-β1 is known to contribute to pain upon bone cancer via upregulation and sensitization of TRPV1 in sensory neurons [189, 190]. In conjunction with these observations, TGFβRI and TGFβRII are known to be upregulated in this rat bone cancer-pain model upon inoculation of rat mammary carcinoma cells [190]. Furthermore, lysophosphatidic acid, LPA, triggers TRPV1 through a PKCε-dependent signaling cascade in dorsal root ganglion neurons upon bone cancer formation in rats [191]. TRPV1 may thus be a central player along the pathways that are behind bone cancer pain in advanced breast cancers.

8. Potential for therapeutical targeting

The emerging role of TRP channels in cancer progression has been widely admitted. Abnormal expression of several TRPV family members, along with the altered expression of other TRP family channels, direct various cancer-linked features, including proliferation, apoptotic control, angiogenesis, and invasive migration leading to distant metastasis [76, 122, 129, 133, 191] (see also **Figure 1**). For that, these calcium channel proteins can also serve as biomarkers and as attractive objectives for therapeutical targeting. The fact that these ion channels can be activated by small pharmacological compounds, also supports their potential for therapeutical approaches and several studies have been performed with potential modulators against the activity of these proteins to target cancer cells [191–195].

TRPV1 channel is activated by a natural compound, capsaicin, the primary pungent component of the chili pepper, and there is evidence for its potential anticancer activity and ability to induce apoptosis [196]. In breast cancer models, ectopic expression of TRPV1 combined with capsaicin-treatment, leads to mitochondrial Ca^{2+} accumulation and necrosis [197]. TRPV1 expression alone was able to stop cell proliferation and induce apoptosis via activation of caspase-3 activity in breast cancer cell lines [197]. As capsaicin, through its impact on TRPV1 activity, also causes pain sensation, it cannot be used as a therapeutical compound in high doses to induce apoptosis. However, a chemical compound, dihydropyridine derivative MRS1477, operates as a modulator of TRPV1 activity, and can be used together with capsaicin to promote apoptosis in breast cancer cells [163]. A study by Wu et al. investigated the mechanisms behind capsaicin-mediated apoptosis by utilizing a TRPV1-inducible MCF-7 breast cancer cell line [162]. They found that the cell death upon capsaicin-treatment was necrotic and linked to elevated levels of c-Fos and receptor-interacting serine/threonine kinase 3, RIP3 that plays a role in the inflammatory mode of cell death, necroptosis [162]. Additionally, an alkyl sulfonamide analogue of capsaicin, RPF151, shows potential for targeting breast cancer cells as shown by studies with

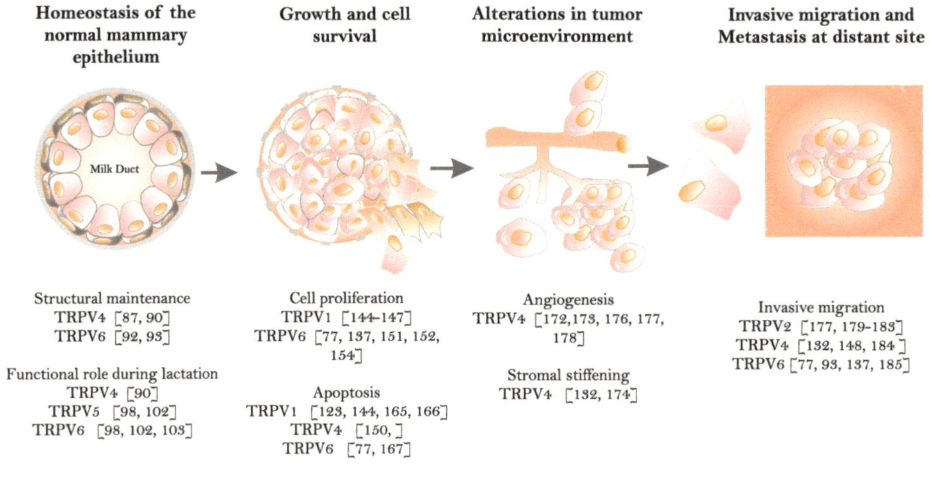

Figure 1.
The role of TRPV family members in the maintenance of normal mammary epithelium and in the induction of hallmarks of cancer. The connection of distinct TRPV family members to the structural and functional maintenance of the normal mammary epithelium as well as their connections to specific steps along breast cancer progression are summarized in this figure. The corresponding references are found within the brackets.

MDA-MB-231 cells [198]. In this study, capsaicin analogue was found to downregulate p21, cyclins A, D1, and D3, subsequently leading to arrest in the S-phase and induction of apoptosis [198]. Furthermore, modulation of TRPV1 activity in sensory neurons by pharmacological compounds may also lead to an anti-tumoral immune response [199]. Systemic treatment with low-dose of capsaicin was shown to trigger an anti-inflammatory response against metastatic breast carcinomas and have potential as a therapy choice [199]. On the other hand, a synthetic antagonist against TRPV1, capsazepine, CPZ, has also been shown to possess anti-cancer effects in vivo through its impact on cell proliferation in several cancer cell types, including breast cancer cells [200]. Capsazepine and its analogues may thus act as potential therapeutic compounds in the future [200].

Besides capsaicin, another natural compound, cannabidiol, has an impact on the induction of apoptosis in MDA-MB-231 breast cancer cells through the TRPV1 channel [201]. These studies showed that besides inducing apoptosis through vanilloid transient receptor potential vanilloid type-1 receptors, cannabidiol can act in the induction of apoptosis via cannabinoid receptor type 2, CB2 and through cannabinoid/vanilloid receptor-independent mechanisms [201]. The interconnection between cannabinoid receptors and TRPV1 has also been investigated in another study that utilized MDA-MB-231 cells as a model. In this study, the role of these receptors in cancer cell invasion was assessed and the results linked activation of both CB1 and TRPV1 by agonist to reduced invasion capability of the MDA-MB-231 cells [202].

Intriguingly, it has also been noticed that some common chemotherapeutic agents interact with TRPV-dependent pathways: The combination of selenium and cisplatin operate through overlapping intracellular pathways and can also modulate TRPV1 activity to induce apoptosis in MCF-7 breast cancer cell line [203]. In addition, combination therapy with alpha-lipoic acid, ALA and cisplatin benefits from the activation of the TRPV1 channel to induce apoptosis in MCF-7 breast cancer cells [204]. Furthermore, in the same breast cancer cell line, chemotherapeutic agent 5-Fluorouracil induces mitochondrial cytotoxicity and apoptosis upon TRPV1 activation [205]. The effectiveness of chemotherapy, combined with the activation of transient receptor potential channel activity, has also been demonstrated with TRPV2: activation of TRPV2 with cannabidiol, CBD, sensitized aggressive triple-negative breast cancer (TNBC) cells to the chemotherapeutic drug, doxorubicin, consequently inhibiting tumor growth in *in vitro* and *in vivo* models [128]. TRPV2 may thus act as a positive prognostic marker for TNBC patients who are undergoing chemotherapy.

Besides induction of apoptosis and inhibition of cell proliferation, TRPV channels have been investigated as potential targets to block invasive migration. TRPV2 has been associated with the function of antimicrobial peptide hCAP18/LL-37, which stimulates both proliferation and migration of various cancer cell types, including breast cancer cells [206]. In line with these previous findings, TRPV2 silencing was found to diminish the LL-37-dependent migration of MCF-7 and MDA-MB-231 breast cancer cells [179]. As TRPV4 is involved in invasive migration as well, and modulation of its activity is possible through several compounds, the potential of targeting this protein should be assessed for reducing metastasis in breast cancer models. Several animal studies have already shown the effectiveness of TRPV4 antagonists as therapeutic agents for treating several other diseases [207, 208]. In addition, TRPV6 is overexpressed in breast cancers and could be targeted in estrogen receptor-negative subtype of mammary carcinomas [136]. Specific TRPV6-targeting compounds have been developed that could be used for manipulating TRPV6 activity and such compounds have also shown promising results in various cancer types, including breast cancer cells [209–212]. While the utilization of TRPV modulators to induce apoptosis or inhibition of either cell proliferation or migration has shown very promising results, one should also consider the risk of other unwanted side-effects through toning of some critical signaling cascades. Such problems can be caused due to the unspecificity of certain antagonists and agonists against the ion channel proteins, leading to the deregulation of several channel protein types. In addition, the wide expression of many of the ion channel proteins throughout various tissues will create challenges in the modulation of ion channel activities at specific sites. For instance, TRPV1 is very widely expressed and most often linked to toxicity in the trials [213, 214]. The balance in the expression and activity of these proteins is though decisive for such a variety of cellular processes.

9. Conclusions

The mammary epithelium is strictly regulated by hormonal signaling, growth factors and cytokines that direct its development, growth and functional organization. In addition, the mammary epithelium is exposed to various physical alterations in the microenvironment that may be sensed by the plasma-membrane embedded structures, such as the ion channels in the mammary epithelial cell populations. Calcium ion pumps and influx through them play a central role in decoding many of the extracellular cues into intracellular signaling. Therefore, these channel proteins greatly impact all essential processes in the maintenance of normal mammary tissue and participate to the development of pathophysiological conditions.

TRPV channels, among the TRP superfamily, are abundantly expressed in discrete tissues, also in the mammary tissue. Of these channel proteins, at least TRPV4-6 have identified functions in the structural and functional maintenance of the normal mammary epithelium, both directly in the mammary epithelium or indirectly through the control of ion influx in other tissues that impact physiological functions of the mammary gland. Whether the other TRPV channels have importance in the structural maintenance of the mammary epithelium or along with lactation, remains to be studied.

Abnormal expression of TRPV channels is also abundantly found in human breast carcinomas and these channel proteins are involved in triggering many of the typical hallmarks of cancers. How TRPV channels are deregulated or aberrantly expressed along breast cancer progression, is still poorly understood. However, as these proteins are sensitive to any physical or biochemical changes in the microenvironment, it is more than likely that they would be affected by the cancer-associated changes in the stroma. This topic certainly requires more investigations in the future. As inducers of the cancer-linked features, such as proliferation, inhibition of apoptosis, invasive migration and angiogenesis, TRPV family members also act as attractive targets for therapeutical choices. A number of known natural and synthetic modulators of TRPV activity already exist and some of them have given promising results in the trials that aim for pharmacological intervention of breast cancers. However, as these proteins are upstream of numerous intracellular pathways that guide cellular functions, there are challenges in such attempts. Furthermore, one should consider the possible interplay in between distinct plasma-membrane embedded calcium channels, several of which may be deregulated along cancer development and impact overlapping intracellular pathways. Such a phenomenon creates a more complex picture on the role of specific ion channels in cancer progression and requires extensive studies in the future.

Author details

Sari Susanna Tojkander
Faculty of Medicine and Health Technology, Tampere Institute for Advanced Study, Tampere University, Finland

*Address all correspondence to: sari.tojkander@tuni.fi

References

[1] Watson CJ, Khaled WT. Mammary development in the embryo and adult: A journey of morphogenesis and commitment. Development. 2008;**135**:995-1003. DOI: 10.1242/dev.005439

[2] Sternlicht MD. Key stages in mammary gland development: The cues that regulate ductal branching morphogenesis. Breast Cancer Research. 2006;**8**:201. DOI: 10.1186/bcr1368

[3] Rios AC, Fu NY, Lindeman GJ, Visvader JE. In situ identification of bipotent stem cells in the mammary gland. Nature. 2014;**506**:322-327. DOI: 10.1038/nature12948

[4] Visvader JE, Stingl J. Mammary stem cells and the differentiation hierarchy: Current status and perspectives. Genes & Development. 2014;**28**:1143-1158. DOI: 10.1101/gad.242511.114

[5] Van Keymeulen A, Rocha A, Ousset M, et al. Distinct stem cells contribute to mammary gland development and maintenance. Nature. 2011;**479**:189-193. DOI: 10.1038/nature10573

[6] Gudjonsson T, Adriance MC, Sternlicht MD, Petersen OW, Bissell MJ. Myoepithelial cells: Their origin and function in breast morphogenesis and neoplasia. Journal of Mammary Gland Biology and Neoplasia. 2005;**10**:261-272. DOI: 10.1007/s10911-005-9586-4

[7] Prater MD, Petit V, Alasdair Russell I, Giraddi RR, Shehata M, Menon S, et al. Mammary stem cells have myoepithelial cell properties. Nature Cell Biology. 2014;**16**:942-950. DOI: 10.1038/ncb3025

[8] Lloyd-Lewis B, Harris OB, Watson CJ, Davis FM. Mammary stem cells: Premise, properties and perspectives. Trends in Cell Biology. 2017;**8**:556-567. DOI: 10.1016/j.tcb.2017.04.001

[9] Faraldo MM, Teulière J, Deugnier MA, Taddei-De La Hosseraye I, Thiery JP, Glukhova MA. Myoepithelial cells in the control of mammary development and tumorigenesis: Data from genetically modified mice. Journal of Mammary Gland Biology and Neoplasia. 2005;**10**:211-219. DOI: 10.1007/s10911-005-9582-8

[10] Moumen M, Chiche A, Cagnet S, Petit V, Raymond K, Faraldo MM, et al. The mammary myoepithelial cell. The International Journal of Developmental Biology. 2011;**55**:763-771. DOI: 10.1387/ijdb.113385mm

[11] Macias H, Hinck L. Mammary gland development. Wiley Interdisciplinary Reviews: Developmental Biology. 2012;**1**:533-557. DOI: 10.1002/wdev.35

[12] Zwick RK, Rudolph MC, Shook BA, Holtrup B, Roth E, Lei V, et al. Adipocyte hypertrophy and lipid dynamics underlie mammary gland remodeling after lactation. Nature Communications. 2018;**9**:3592. DOI: 10.1038/s41467-018-05911-0

[13] Lloyd-Lewis B, Davis FM, Harris OB, Hitchcock JR, Watson CJ. Neutral lineage tracing of proliferative embryonic and adult mammary stem/progenitor cells. Development. 2018;**145**:dev164079. DOI: 10.1242/dev.164079

[14] Davis FM. The ins and outs of calcium signalling in lactation and involution: Implications for breast cancer treatment. Pharmacological Research. 2017;**116**:100-104. DOI: 10.1016/j.phrs.2016.12.007

[15] McManaman JL, Neville MC. Mammary physiology and milk secretion. Advanced Drug Delivery Reviews. 2003;**55**:629-641. DOI: 10.1016/s0169-409x(03)00033-4

[16] Haaksma CJ, Schwartz RJ, Tomasek JJ. Myoepithelial cell contraction and milk ejection are impaired in mammary glands of mice lacking smooth muscle alpha-actin. Biology of Reproduction. 2011;**85**:13-21. DOI: 10.1095/biolreprod.110.090639

[17] Raymond K, Cagnet S, Kreft M, Janssen H, Sonnenberg A, Glukhova MA. Control of mammary myoepithelial cell contractile function by α3β1 integrin signalling. The EMBO Journal. 2011;**30**:1896-1906. DOI: 10.1038/emboj.2011.113

[18] Kreuzaler PA, Staniszewska AD, Li W, Omidvar N, Kedjouar B, Turkson J, et al. Stat3 controls lysosomal-mediated cell death in vivo. Nature Cell Biology. 2011;**13**:303-309. DOI: 10.1038/ncb2171

[19] Jindal S, Gao D, Bell P, Albrektsen G, Edgerton SM, Ambrosone CB, et al. Postpartum breast involution reveals regression of secretory lobules mediated by tissue-remodeling. Breast Cancer Research. 2014;**16**:R31. DOI: 10.1186/bcr3633

[20] Sargeant TJ, Lloyd-Lewis B, Resemann HK, Ramos-Montoya A, Skepper J, Watson CJ. Stat3 controls cell death during mammary gland involution by regulating uptake of milk fat globules and lysosomal membrane permeabilization. Nature Cell Biology. 2014;**16**:1057-1068. DOI: 10.1038/ncb3043

[21] Brown RC, Davis TP. Calcium modulation of adherens and tight junction function: A potential mechanism for blood-brain barrier disruption after stroke. Stroke. 2002;**33**:1706-1711. DOI: 10.1161/01.str.0000016405.06729.83

[22] Islam MS. Calcium signaling: From basic to bedside. Advances in Experimental Medicine and Biology. 2020;**1131**:1-6. DOI: 10.1007/978-3-030-12457-1_1

[23] Venkatachalam K, Montell C. Trp channels. Annual Review of Biochemistry. 2007;**76**:387-417. DOI: 10.1146/annurev.biochem.75.103004.142819

[24] Volkers L, Mechioukhi Y, Coste B. Piezo channels: From structure to function. Pflügers Archiv. 2015;**467**:95-99. DOI: 10.1007/s00424-014-1578-z

[25] Stewart TA, Yapa KT, Monteith GR. Altered calcium signaling in cancer cells. Biochimica et Biophysica Acta. 2015;**1848**:2502-2511. DOI: 10.1016/j.bbamem.2014.08.016

[26] So CL, Saunus JM, Roberts-Thomson SJ, Monteith GR. Calcium signalling and breast cancer. Seminars in Cell & Developmental Biology. 2019;**94**:74-83. DOI: 10.1016/j.semcdb.2018.11.001

[27] Cross BM, Breitwieser GE, Reinhardt TA, Rao R. Cellular calcium dynamics in lactation and breast cancer: From physiology to pathology. American Journal of Physiology. Cell Physiology. 2014;**306**:C515-C526. DOI: 10.1152/ajpcell.00330.2013

[28] Leclercq G. Calcium-induced activation of estrogen receptor alpha—New insight. Steroids. 2012;**77**:924-927. DOI: 10.1016/j.steroids.2012.01.012

[29] Vanhouten JN, Wysolmerski JJ. The calcium-sensing receptor in the breast.

Best Practice & Research. Clinical Endocrinology & Metabolism. 2013;**27**:403-414. DOI: 10.1016/j.beem.2013.02.011

[30] Jurek B, Neumann ID. The oxytocin receptor: From intracellular signaling to behavior. Physiological Reviews. 2018;**98**:1805-1908. DOI: 10.1152/physrev.00031.2017

[31] Montell C, Birnbaumer L, Flockerzi V. The TRP channels, a remarkably functional family. Cell. 2002;**108**:595-598. DOI: 10.1016/s0092-8674(02)00670-0

[32] Vangeel L, Voets T. Transient receptor potential channels and calcium signaling. Cold Spring Harbor Perspectives in Biology. 2019;**11**:a035048. DOI: 10.1101/cshperspect.a035048

[33] Tsagareli MG, Nozadze I. An overview on transient receptor potential channels superfamily. Behavioural Pharmacology. 2020;**31**:413-434. DOI: 10.1097/FBP.0000000000000524

[34] Garcia-Sanz N. Identification of a tetramerization domain in the C terminus of the vanilloid receptor. The Journal of Neuroscience. 2004;**24**:5307-5314. DOI: 10.1523/JNEUROSCI.0202-04.2004

[35] Smith GD, Gunthorpe MJ, Kelsell RE, Hayes PD, Reilly P, Facer P, et al. TRPV3 is a temperature-sensitive vanilloid receptor-like protein. Nature. 2002;**418**:186-190. DOI: 10.1007/978-981-10-7757-9_6

[36] Hellwig N, Albrecht N, Harteneck C, Schultz G, Schaefer M. Homo- and heteromeric assembly of TRPV channel subunits. Journal of Cell Science. 2005;**118**:917-928. DOI: 10.1242/jcs.01675

[37] Plant TD, Strotmann R. TRPV4. Handbook of Experimental Pharmacology. 2007;**179**:189-205. DOI: 10.1007/978-3-540-34891-7_11

[38] Peng JB, Chen XZ, Berger UV, Vassilev PM, Tsukaguchi H, Brown EM, et al. Molecular cloning and characterization of a channel-like transporter mediating intestinal calcium absorption. The Journal of Biological Chemistry. 1999;**274**:22739-22746. DOI: 10.1074/jbc.274.32.22739

[39] Wissenbach U, Niemeyer BA, Fixemer T, Schneidewind A, Trost C, Cavalie A, et al. Expression of CaT-like, a novel calcium-selective channel, correlates with the malignancy of prostate cancer. The Journal of Biological Chemistry. 2001;**276**:19461-19468. DOI: 10.1074/jbc.M009895200

[40] Voets T, Janssens A, Prenen J, Droogmans G, Nilius B. Mg^{2+}-dependent gating and strong inward rectification of the cation channel TRPV6. The Journal of General Physiology. 2003;**121**:245-260. DOI: 10.1085/jgp.20028752

[41] Clapham DE. TRP channels as cellular sensors. Nature. 2003;**426**:517-524. DOI: 10.1038/nature02196

[42] Mutai H, Heller S. Vertebrate and invertebrate TRPV-like mechanoreceptors. Cell Calcium. 2003;**33**:471-478. DOI: 10.1016/s0143-4160(03)00062-9

[43] Caterina MJ, Rosen TA, Tominaga M, Brake AJ, Julius DA. Capsaicin-receptor homologue with a high threshold for noxious heat. Nature. 1999;**398**:436-441. DOI: 10.1038/18906

[44] Caterina MJ, Schumacher MA, Tominaga M, Rosen TA, Levine JD, Julius D. The capsaicin receptor: A heat-activated ion channel in the pain pathway. Nature. 1997;**389**:816-824. DOI: 10.1038/39807

[45] Tominaga M, Caterina MJ. Thermosensation and pain. Journal of Neurobiology. 2004;**61**:3-12. DOI: 10.1002/neu.20079

[46] Caterina MJ. Impaired nociception and pain sensation in mice lacking the capsaicin receptor. Science. 2000;**288**:306-313. DOI: 10.1126/science.288.5464.306

[47] Shibasaki K, Murayama N, Ono K, Ishizaki Y, Tominaga M. TRPV2 enhances axon outgrowth through its activation by membrane stretch in developing sensory and motor neurons. The Journal of Neuroscience. 2010;**30**:4601-4612. DOI: 10.1523/JNEUROSCI.5830-09.2010

[48] Perálvarez-Marín A, Doñate-Macian P, Gaudet R. What do we know about the transient receptor potential vanilloid 2 (TRPV2) ion channel? The FEBS Journal. 2013;**280**:5471-5487. DOI: 10.1111/febs.12302

[49] Cohen MR, Johnson WM, Pilat JM, Kiselar J, DeFrancesco-Lisowitz A, Zigmond RE, et al. Nerve growth factor regulates transient receptor potential vanilloid 2 via extracellular signal-regulated kinase signaling to enhance neurite outgrowth in developing neurons. Molecular and Cellular Biology. 2015;**35**:4238-4252. DOI: 10.1128/MCB.00549-15

[50] Mercado J, Gordon-Shaag A, Zagotta WN, Gordon SE. Ca2+-dependent desensitization of TRPV2 channels is mediated by hydrolysis of phosphatidylinositol 4,5-bisphosphate. The Journal of Neuroscience. 2010;**30**:13338-13347. DOI: 10.1523/JNEUROSCI.2108-10.2010

[51] Liberati S, Morelli M, Amantini C, Farfariello V, Santoni M, Conti A, et al. Loss of TRPV2 homeostatic control of cell proliferation drives tumor progression. Cell. 2014;**3**:112-128. DOI: 10.3390/cells3010112

[52] Shibasaki K. Physiological significance of TRPV2 as a mechanosensor, thermosensor and lipid sensor. The Journal of Physiological Sciences. 2016;**66**:359-365. DOI: 10.1007/s12576-016-0434-7

[53] Peier AM. A heat-sensitive TRP channel expressed in keratinocytes. Science. 2002;**296**:2046-2049. DOI: 10.1126/science.1073140

[54] Caterina MJ, Pang Z. TRP channels in skin biology and pathophysiology. Pharmaceuticals (Basel). 2016;**9**:77. DOI: 10.3390/ph9040077

[55] Yamamoto-Kasai E, Imura K, Yasui K, Shichijou M, Oshima I, Hirasawa T, et al. TRPV3 as a therapeutic target for itch. The Journal of Investigative Dermatology. 2012;**132**:2109-2112. DOI: 10.1038/jid.2012.97

[56] Strotmann R, Harteneck C, Nunnenmacher K, Schultz G, Plant TD. OTRPC4, a nonselective cation channel that confers sensitivity to extracellular osmolarity. Nature Cell Biology. 2000;**2**:695-702. DOI: 10.1038/35036318

[57] Gao X, Wu L, O'Neil RG. Temperature-modulated diversity of TRPV4 channel gating: Activation by physical stresses and phorbol ester derivatives through protein kinase C-dependent and -independent pathways. The Journal of Biological Chemistry. 2003;**278**:27129-27137. DOI: 10.1074/jbc.M302517200

[58] Liedtke W, Friedman JM. Abnormal osmotic regulation in trpv4−/− mice. Proceedings of the National Academy of Sciences of the United States of America.

2003;**100**:13698-13703. DOI: 10.1073/pnas.1735416100

[59] Suzuki M, Mizuno A, Kodaira K, Imai M. Impaired pressure sensation in mice lacking TRPV4. The Journal of Biological Chemistry. 2003;**278**:22664-22668. DOI: 10.1074/jbc.M302561200

[60] Nilius B, Vriens J, Prenen J, Droogmans G, Voets T. TRPV4 calcium entry channel: A paradigm for gating diversity. American Journal of Physiology. Cell Physiology. 2004;**286**:C195-C205. DOI: 10.1152/ajpcell.00365.2003

[61] Shibasaki K. TRPV4 ion channel as important cell sensors. Journal of Anesthesia. 2016;**30**:1014-1019. DOI: 10.1007/s00540-016-2225-y

[62] White JPM, Cibelli M, Urban L, Nilius B, McGeown JG, Nagy I. TRPV4: Molecular conductor of a diverse orchestra. Physiological Reviews. 2016;**96**:911-973. DOI: 10.1152/physrev.00016.2015

[63] Hoenderop JG, Nilius B, Bindels RJ. Epithelial calcium channels: From identification to function and regulation. Pflügers Archiv. 2003;**446**:304-308. DOI: 10.1007/s00424-003-1045-8

[64] Nijenhuis T, Hoenderop JG, Nilius B, Bindels RJ. (Patho)physiological implications of the novel epithelial Ca2+ channels TRPV5 and TRPV6. Pflügers Archiv. 2003;**446**:401-409. DOI: 10.1007/s00424-003-1038-7

[65] van Abel M, Hoenderop JG, Bindels RJ. The epithelial calcium channels TRPV5 and TRPV6: Regulation and implications for disease. Naunyn-Schmiedeberg's Archives of Pharmacology. 2005;**371**:295-306. DOI: 10.1007/s00210-005-1021-2

[66] Bianco SD, Peng JB, Takanaga H, Suzuki Y, Crescenzi A, Kos CH, et al. Marked disturbance of calcium homeostasis in mice with targeted disruption of the Trpv6 calcium channel gene. Journal of Bone and Mineral Research. 2007;**22**:274-285. DOI: 10.1359/jbmr.061110

[67] Saotome K, Singh AK, Sobolevsky AI. Determining the crystal structure of TRPV6. In: Kozak JA, Putney JW Jr, editors. Calcium Entry Channels in Non-Excitable Cells. Boca Raton (FL): CRC Press/Taylor & Francis; 2018. Chapter 14. DOI: 10.1201/9781315152592-14

[68] Meyer MB, Watanuki M, Kim S, Shevde NK, Pike JW. The human transient receptor potential vanilloid type 6 distal promoter contains multiple vitamin D receptor binding sites that mediate activation by 1,25-dihydroxyvitamin D3 in intestinal cells. Molecular Endocrinology. 2006;**20**:1447-1461. DOI: 10.1210/me.2006-0031

[69] Wissenbach U, Niemeyer BA. TRPV6. Handbook of Experimental Pharmacology. 2007;**179**:221-234. DOI: 10.1007/978-3-540-34891-7_13

[70] Balesaria S, Sangha S, Walters JR. Human duodenum responses to vitamin D metabolites of TRPV6 and other genes involved in calcium absorption. American Journal of Physiology—Gastrointestinal and Liver Physiology. 2009;**297**:G1193-G1197. DOI: 10.1152/ajpgi.00237.2009

[71] Na T, Peng JB. TRPV5: A Ca(2+) channel for the fine-tuning of Ca(2+) reabsorption. Handbook of Experimental Pharmacology. 2014;**222**:321-357. DOI: 10.1007/978-3-642-54215-2_13

[72] Fecher-Trost C, Weissgerber P, Wissenbach U. TRPV6 channels.

Handbook of Experimental Pharmacology. 2014;**222**:359-384. DOI: 10.1007/978-3-642-54215-2_14

[73] Fecher-Trost C, Wissenbach U, Weissgerber P. TRPV6: From identification to function. Cell Calcium. 2017;**67**:116-122. DOI: 10.1016/j.ceca.2017.04.006

[74] Bahamonde MI, Fernandez-Fernandez JM, Guix FX, Vazquez E, Valverde MA. Plasma membrane voltage-dependent anion channel mediates antiestrogen-activated maxi Cl- currents in C1300 neuroblastoma cells. The Journal of Biological Chemistry. 2003;**278**:33284-33289. DOI: 10.1074/jbc.M302814200

[75] Irnaten M, Blanchard-Gutton N, Praetorius J, Harvey BJ. Rapid effects of 17beta-estradiol on TRPV5 epithelial Ca(2+) channels in rat renal cells. Steroids. 2009;**74**:642-649. DOI: 10.1016/j.ceca.2008.02.007

[76] Bolanz KA, Hediger MA, Landowski CP. The role of TRPV6 in breast carcinogenesis. Molecular Cancer Therapeutics. 2008;**7**:271-279. DOI: 10.1158/1535-7163.MCT-07-0478

[77] Arendt LM, Kuperwasser C. Form and function: How estrogen and progesterone regulate the mammary epithelial hierarchy. Journal of Mammary Gland Biology and Neoplasia. 2015;**20**:9-25. DOI: 10.1007/s10911-015-9337-0

[78] Jung C, Fandos C, Lorenzo IM, Plata C, Fernandes J, Gené GG, et al. The progesterone receptor regulates the expression of TRPV4 channel. Pflügers Archiv. 2009;**459**:105-113. DOI: 10.1007/s00424-009-0706-7

[79] Russo J, Ao X, Grill C, Russo IH. Pattern of distribution of cells positive for estrogen receptor alpha and progesterone receptor in relation to proliferating cells in the mammary gland. Breast Cancer Research and Treatment. 1999;**53**:217-227. DOI: 10.1023/a:1006186719322

[80] Schultz JR, Petz LN, Nardulli AM. Estrogen receptor alpha and Sp1 regulate progesterone receptor gene expression. Molecular and Cellular Endocrinology. 2003;**201**:165-175. DOI: 10.1016/s0303-7207(02)00415-x

[81] Petz LN, Ziegler YS, Schultz JR, Kim H, Kemper JK, Nardulli AM. Differential regulation of the human progesterone receptor gene through an estrogen response element half site and Sp1 sites. The Journal of Steroid Biochemistry and Molecular Biology. 2004;**88**:113-122. DOI: 10.1016/j.jsbmb.2003.11.008

[82] Lee WJ, Monteith GR, Roberts-Thomson SJ. Calcium transport and signaling in the mammary gland: Targets for breast cancer. Biochimica et Biophysica Acta. 2006;**1765**:235-255. DOI: 10.1016/j.bbcan.2005.12.001

[83] Sokabe T, Fukumi-Tominaga T, Yonemura S, Mizuno A, Tominaga M. The TRPV4 channel contributes to intercellular junction formation in keratinocytes. The Journal of Biological Chemistry. 2010;**285**:18749-18758. DOI: 10.1074/jbc.M110.103606

[84] Akazawa Y, Yuki T, Yoshida H, Sugiyama Y, Inoue S. Activation of TRPV4 strengthens the tight-junction barrier in human epidermal keratinocytes. Skin Pharmacology and Physiology. 2013;**26**:15-21. DOI: 10.1159/000343173

[85] Janssen DA, Jansen CJ, Hafmans TG, Verhaegh GW, Hoenderop JG, Heesakkers JP, et al. TRPV4 channels in the human urogenital

tract play a role in cell junction formation and epithelial barrier. Acta Physiologica (Oxford, England). 2016;**218**:38-48. DOI: 10.1111/apha.12701

[86] Reiter B, Kraft R, Günzel D, Zeissig S, Schulzke JD, Fromm M, et al. TRPV4-mediated regulation of epithelial permeability. The FASEB Journal. 2006;**20**:1802-1812. DOI: 10.1096/fj.06-5772com

[87] Martínez-Rendón J, Sánchez-Guzmán E, Rueda A, González J, Gulias-Cañizo R, Aquino-Jarquín G, et al. TRPV4 regulates tight junctions and affects differentiation in a cell culture model of the corneal epithelium. Journal of Cellular Physiology. 2017;**232**:1794-1807. DOI: 10.1002/jcp.25698

[88] Mukaiyama M, Yamasaki Y, Usui T, Nagumo Y. Transient receptor potential V4 channel stimulation induces reversible epithelial cell permeability in MDCK cell monolayers. FEBS Letters. 2019;**593**:2250-2260. DOI: 10.1002/1873-3468.13490

[89] Islam MA, Mizusawa M, Sharmin MM, Hayashi S, Yonekura S. TRPV4 increases the expression of tight junction protein-encoding genes via XBP1 in mammary epithelial cells. Animals. 2020;**10**:1174. DOI: 10.3390/ani10071174

[90] Kitsuki T, Yoshimoto RU, Aijima R, Hatakeyama J, Cao AL, Zhang JQ, et al. Enhanced junctional epithelial permeability in TRPV4-deficient mice. Journal of Periodontal Research. 2020;**55**:51-60. DOI: 10.1111/jre.12685

[91] Zhao D, Han X, Huang L, Wang J, Zhang X, Jeon JH, et al. Transcription factor ZFHX3 regulates calcium influx in mammary epithelial cells in part via the TRPV6 calcium channel. Biochemical and Biophysical Research Communications. 2019;**519**:366-371. DOI: 10.1016/j.bbrc.2019.08.148

[92] Kärki T, Rajakylä EK, Acheva A, Tojkander S. TRPV6 calcium channel directs homeostasis of the mammary epithelial sheets and controls epithelial mesenchymal transition. Scientific Reports. 2020;**10**:14683. DOI: 10.1038/s41598-020-71645-z

[93] Hennighausen L, Robinson GW. Information networks in the mammary gland. Nature Reviews. Molecular Cell Biology. 2005;**6**:715-725. DOI: 10.1038/nrm1714

[94] Brisken C, O'Malley B. Hormone action in the mammary gland. Cold Spring Harbor Perspectives in Biology. 2010;**2**:a003178. DOI: 10.1101/cshperspect.a003178

[95] Sumbal J, Chiche A, Charifou E, Koledova Z, Li H. Primary mammary organoid model of lactation and involution. Frontiers in Cell and Development Biology. 2020;**9**(8):68. DOI: 10.3389/fcell.2020.00068

[96] Kovacs CS. Calcium and phosphate metabolism and related disorders during pregnancy and lactation. 2021. In: Feingold KR, Anawalt B, Boyce A, Chrousos G, de Herder WW, Dhatariya K, Dungan K, et al. editors. Endotext [Internet]. South Dartmouth (MA): MDText.com, Inc.; 2000

[97] Beggs MR, Appel I, Svenningsen P, Skjødt K, Alexander RT, Dimke H. Expression of transcellular and paracellular calcium and magnesium transport proteins in renal and intestinal epithelia during lactation. American Journal of Physiology. Renal Physiology. 2017;**313**:F629-F640. DOI: 10.1152/ajprenal.00680.2016

[98] Hoenderop JGJ, van der Kemp AWCM, Hartog A, van de Graaf SFJ, van Os CH, Willems PHGM, et al. Molecular identification of the apical Ca2+channel in 1,25-dihydroxyvitamin D3-responsive epithelia. The Journal of Biological Chemistry. 1999;**274**:8375-8378. DOI: 10.1074/jbc.274.13.8375

[99] Hoenderop JG, Nilius B, Bindels RJ. Molecular mechanism of active Ca2+ reabsorption in the distal nephron. Annual Review of Physiology. 2002;**64**:529-549. DOI: 10.1146/annurev.physiol.64.081501.155921

[100] Peng JB, Suzuki Y, Gyimesi G, Hediger MA. TRPV5 and TRPV6 calcium-selective channels. In: Kozak JA, Putney JW, editors. Calcium Entry Channels in Non-Excitable Cells, Methods in Signal Transduction Series. Boca Raton, FL, USA: CRC Press/Taylor & Francis; 2017. pp. 241-274. DOI: 10.1201/9781315152592-13. ISBN 978-1-315-15259-2

[101] Hoenderop JGJ, Nilius B, Bindels RJM. Calcium absorption across epithelia. Physiological Reviews. 2005;**85**:373-422. DOI: 10.1152/physrev.00003.2004

[102] Ajibade DV, Dhawan P, Fechner AJ, Meyer MB, Pike JW, Christakos S. Evidence for a role of prolactin in calcium homeostasis: Regulation of intestinal transient receptor potential vanilloid type 6, intestinal calcium absorption, and the 25-hydroxyvitamin D(3) 1alpha hydroxylase gene by prolactin. Endocrinology. 2010;**151**:2974-2984. DOI: 10.1210/en.2010-0033

[103] Kobayashi K, Tsugami Y, Matsunaga K, Suzuki T, Nishimura T. Moderate high temperature condition induces the lactation capacity of mammary epithelial cells through control of stat3 and stat5 signaling. Journal of Mammary Gland Biology and Neoplasia. 2018;**23**:75-88. DOI: 10.1007/s10911-018-9393-3

[104] Mizusawa M, Sharmin MM, Yonekura S. Mild heat stress induces transcription of the beta-casein gene via unfolded protein response-activated xbp1 signaling in undifferentiated mammary epithelial cells. Animal Science Journal. 2019;**90**:1026-1032. DOI: 10.1111/asj.13246

[105] Park HG, Han SI, Oh SY, Kang HS. Cellular responses to mild heat stress. Cellular and Molecular Life Sciences. 2005;**62**:10-23. DOI: 10.1007/s00018-004-4208-7

[106] Guler AD, Lee H, Iida T, Shimizu I, Tominaga M, Caterina M. Heat-evoked activation of the ion channel, trpv4. The Journal of Neuroscience. 2002;**22**:6408-6414. DOI: 10.1523/JNEUROSCI.22-15-06408.2002

[107] Jordt SE, McKemy DD, Julius D. Lessons from peppers and peppermint: The molecular logic of thermosensation. Current Opinion in Neurobiology. 2003;**13**:487-492. DOI: 10.1016/s0959-4388(03)00101-6

[108] Patapoutian A, Peier AM, Story GM, Viswanath V. Thermo TRP channels and beyond: Mechanisms of temperature sensation. Nature Reviews. Neuroscience. 2003;**4**:529-539. DOI: 10.1038/nrn1141

[109] Nguyen DA, Neville MC. Tight junction regulation in the mammary gland. Journal of Mammary Gland Biology and Neoplasia. 1998;**3**:233-246. DOI: 10.1023/a:1018707309361

[110] Xu X, Gupta S, Hu W, McGrath BC, Cavener DR. Hyperthermia induces the er stress pathway. PLoS One. 2011;**6**:e23740. DOI: 10.1371/journal.pone.0023740

[111] Tsuchiya M, Koizumi Y, Hayashi S, Hanaoka M, Tokutake Y, Yonekura S. The role of unfolded protein response in differentiation of mammary epithelial cells. Biochemical and Biophysical Research Communications. 2017;**484**:903-908. DOI: 10.1016/j.bbrc.2017.02.042

[112] Wainwright A, Rutter AR, Seabrook GR, Reilly K, Oliver KR. Discrete expression of TRPV2 within the hypothalamo-neurohypophysial system: Implications for regulatory activity within the hypothalamic-pituitary-adrenal axis. The Journal of Comparative Neurology. 2004;**474**:24-42. DOI: 10.7599/hmr.2011.31.2.116

[113] Watson CJ, Kreuzaler PA. Remodeling mechanisms of the mammary gland during involution. The International Journal of Developmental Biology. 2011;**55**:757-762. DOI: 10.1387/ijdb.113414cw

[114] Li J, Chen Z, Su K, Zeng J. Clinicopathological classification and traditional prognostic indicators of breast cancer. International Journal of Clinical and Experimental Pathology. 2015;**8**:8500-8505

[115] Ouadid-Ahidouch H, Dhennin-Duthille I, Gautier M, Sevestre H, Ahidouch A. Canaux cationiques TRP dans le cancer du sein: Expression, rôle et corrélation avec des paramètres cliniques [TRP calcium channel and breast cancer: Expression, role and correlation with clinical parameters]. Bulletin du Cancer. 2012;**99**:655-664. French. DOI: 10.1684/bdc.2012.1595

[116] Yang D, Kim J. Emerging role of transient receptor potential (TRP) channels in cancer progression. BMB Reports. 2020;**53**:125-132. DOI: 10.5483/BMBRep.2020.53.3.016

[117] Hanahan D, Weinberg RA. Hallmarks of cancer: The next generation. Cell. 2011;**144**:646-674. DOI: 10.1016/j.cell.2011.02.013

[118] Déliot N, Constantin B. Plasma membrane calcium channels in cancer: Alterations and consequences for cell proliferation and migration. Biochimica et Biophysica Acta. 2015;**1848**(10 Pt B):2512-2522. DOI: 10.1016/j.bbamem.2015.06.009

[119] Rokhlin O, Taghiyev AF, Bayer KU, Bumcrot D, Kotelianski VE, Glover RA, et al. Calcium/calmodulin-dependent kinase II plays an important role in prostate cancer cell survival. Cancer Biology & Therapy. 2007;**6**:732-742. DOI: 10.4161/cbt.6.5.3975

[120] Shapovalov G, Ritaine A, Skryma R, Prevarskaya N. Role of TRP ion channels in cancer and tumorigenesis. Seminars in Immunopathology. 2016;**38**:357-369. DOI: 10.1007/s00281-015-0525-1

[121] Weber LV, Al-Refae K, Wolk G, Bonatz G, Altmuller J, Becker C, et al. Expression and functionality of TRPV1 in breast cancer cells. Breast Cancer. 2016;**8**:243-252. DOI: 10.2147/BCTT.S121610

[122] So CL, Milevskiy MJG, Monteith GR. Transient receptor potential cation channel subfamily V and breast cancer. Laboratory Investigation. 2020;**100**:199-206. DOI: 10.1038/s41374-019-0348-0

[123] Curtis C, Shah SP, Chin SF, Turashvili G, Rueda OM, Dunning MJ, et al. The genomic and transcriptomic

architecture of 2,000 breast tumours reveals novel subgroups. Nature. 2012;**486**:346-352. DOI: 10.1038/nature10983

[124] Pereira B, Chin SF, Rueda OM, Vollan HK, Provenzano E, Bardwell HA, et al. The somatic mutation profiles of 2,433 breast cancers refines their genomic and transcriptomic landscapes. Nature Communications. 2016;**7**:11479. DOI: 10.1038/ncomms11479

[125] Lozano C, Cordova C, Marchant I, Zuniga R, Ochova P, Ramirez-Barrantes R, et al. Intracellular aggregated TRPV1 is associated with lower survival in breast cancer patients. Breast Cancer. 2018;**10**:161-168. DOI: 10.2147/BCTT.S170208

[126] Santoni G, Amantini C, Maggi F, Marinelli O, Santoni M, Nabissi M, et al. The TRPV2 cation channels: From urothelial cancer invasiveness to glioblastoma multiforme interactome signature. Laboratory Investigation. 2020;**100**:186-198. DOI: 10.1038/s41374-019-0333-7

[127] Siveen KS, Nizamuddin PB, Uddin S, Al-Thani M, Frenneaux MP, Janahi IA, et al. TRPV2: A cancer biomarker and potential therapeutic target. Disease Markers. 2020;**2020**:8892312. DOI: 10.1155/2020/8892312

[128] Elbaz M, Ahirwar D, Xiaoli Z, Zhou X, Lustberg M, Nasser MW, et al. TRPV2 is a novel biomarker and therapeutic target in triple negative breast cancer. Oncotarget. 2018;**9**:33459-33470. DOI: 10.18632/oncotarget.9663

[129] Bahari NN, Jamaludin SYN, Jahidin AH, Zahary MN, Hilmi ABM, Bakar NHA, et al. The emerging roles of TRPV4 in cancer. Biomedical and Pharmacology Journal. 2017;**10**(4):1757-1764. DOI: 10.13005/bpj/1289

[130] Lee WH, Choong LY, Jin TH, Mon NN, Chong S, Liew CS, et al. TRPV4 plays a role in breast cancer cell migration via Ca2+–dependent activation of AKT and downregulation of E-cadherin cell cortex protein. Oncogene. 2017;**22**(6):e338. DOI: 10.1038/oncsis.2017.39

[131] Peng JB, Zhuang L, Berger UV, Adam RM, Williams BJ, Brown EM, et al. CaT1 expression correlates with tumor grade in prostate cancer. Biochemical and Biophysical Research Communications. 2001;**282**:729-734. DOI: 10.1006/bbrc.2001.4638

[132] Fixemer T, Wissenbach U, Flockerzi V, Bonkhoff H. Expression of the Ca2+–selective cation channel TRPV6 in human prostate cancer: A novel prognostic marker for tumor progression. Oncogene. 2003;**22**:7858-7861. DOI: 10.1038/sj.onc.1206895

[133] Lehen'kyi V, Raphael M, Prevarskaya N. The role of the TRPV6 channel in cancer. The Journal of Physiology. 2012;**590**:1369-1376. DOI: 10.1113/jphysiol.2011.225862

[134] Zhuang LY, Peng JB, Tou LQ, Takanaga H, Adam RM, Hediger MA, et al. Calcium-selective ion channel, CaT1, is apically localized in gastrointestinal tract epithelia and is aberrantly expressed in human malignancies. Laboratory Investigation. 2002;**82**:1755-1764. DOI: 10.1097/01.lab.0000043910.41414.e7

[135] Dhennin-Duthille I, Gautier M, Faouzi M, Guilbert A, Brevet M, Vaudry D, et al. High expression of transient receptor potential channels in human breast cancer epithelial cells and tissues correlation with pathological parameters. Cellular Physiology and Biochemistry. 2011;**28**:813-822. DOI: 10.1159/000335795

[136] Peters AA, Simpson PT, Bassett JJ, Lee JM, Silva LD, Reid LE, et al. Calcium channel TRPV6 as a potential therapeutic target in estrogen receptor-negative breast cancer. Molecular Cancer Therapeutics. 2012;**11**:2158-2168. DOI: 10.1158/1535-7163.MCT-11-0965

[137] Roderick HL, Cook SJ. Ca2+ signalling checkpoints in cancer: Remodelling Ca2+ for cancer cell proliferation and survival. Nature Reviews. Cancer. 2008;**8**:361-375. DOI: 10.1038/nrc2374

[138] Chen YF, Chen YT, Chiu WT, Shen MR. Remodeling of calcium signaling in tumor progression. Journal of Biomedical Science. 2013;**20**:23. DOI: 10.1186/1423-0127-20-23

[139] Monteith GR, Davis FM, Roberts-Thomson SJ. Calcium channels and pumps in cancer: Changes and consequences. The Journal of Biological Chemistry. 2012;**287**:31666-31673. DOI: 10.1074/jbc.R112.343061

[140] Monteith GR, Prevarskaya N, Roberts-Thomson SJ. The calcium-cancer signalling nexus. Nature Reviews. Cancer. 2017;**17**:367-380. DOI: 10.1038/nrc.2017.18

[141] Zhai K, Liskova A, Kubatka P, Büsselberg D. Calcium entry through TRPV1: A potential target for the regulation of proliferation and apoptosis in cancerous and healthy cells. International Journal of Molecular Sciences. 2020;**21**:4177. DOI: 10.3390/ijms21114177

[142] Vercelli C, Barbero R, Cuniberti B, Racca S, Abbadessa G, et al. Transient receptor potential vanilloid 1 expression and functionality in mcf-7 cells: A preliminary investigation. Journal of Breast Cancer. 2014;**17**:332-338. DOI: 10.4048/jbc.2014.17.4.332

[143] Kojima I, Nagasawa M. TRPV2: A calcium-permeable cation channel regulated by insulin-like growth factors. In: Liedtke WB, Heller S, editors. TRP Ion Channel Function in Sensory Transduction and Cellular Signaling Cascades. Boca Raton (FL): CRC Press/Taylor & Francis; 2007. Chapter 7

[144] Nie L, Oishi Y, Doi I, Shibata H, Kojima I. Inhibition of proliferation of MCF-7 breast cancer cells by a blocker of Ca2+–permeable channel. Cell Calcium. 1997;**22**:75-82. DOI: 10.1016/s0143-4160(97)90107-x

[145] Lee WH, Choong LY, Mon NN, Lu S, Lin Q, Pang B, et al. TRPV4 regulates breast cancer cell extravasation, stiffness and actin cortex. Scientific Reports. 2016;**6**:27903. DOI: 10.1038/srep27903

[146] Yu S, Huang S, Ding Y, et al. Transient receptor potential ion-channel subfamily V member 4: A potential target for cancer treatment. Cell Death & Disease. 2019;**10**:497. DOI: 10.1038/s41419-019-1708-9

[147] Peters AA, Jamaludin SYN, Yapa KTDS, Chalmers S, Wiegmans AP, Lim HF, et al. Oncosis and apoptosis induction by activation of an overexpressed ion channel in breast cancer cells. Oncogene. 2017;**36**:6490-6500. DOI: 10.1038/onc.2017.234

[148] Kaemmerer E, Turner D, Peters AA, Roberts-Thomson SJ, Monteith GR. An automated epifluorescence microscopy imaging assay for the identification of phospho-AKT level modulators in breast cancer cells. Journal of Pharmacological and Toxicological Methods. 2018;**92**:13-19. DOI: 10.1016/j.vascn.2018.02.005

[149] Cai R, Wang L, Liu X, Michalak M, Tang J, Peng JB, et al. Auto-inhibitory intramolecular S5/S6 interaction in the TRPV6 channel regulates breast

cancer cell migration and invasion. Communications Biology. 2021;**4**:990. DOI: 10.1038/s42003-021-02521-3

[150] Bolanz KA, Kovacs GG, Landowski CP, Hediger MA. Tamoxifen inhibits TRPV6 activity via estrogen receptor-independent pathways in TRPV6-expressing MCF-7 breast cancer cells. Molecular Cancer Research. 2009;**7**:2000-2010. DOI: 10.1158/1541-7786.MCR-09-0188

[151] Kim SY, Yang D, Myeong J, Ha K, Kim SH, Park EJ, et al. Regulation of calcium influx and signaling pathway in cancer cells via TRPV6-Numb1 interaction. Cell Calcium. 2013;**53**:102-111. DOI: 10.1016/j.ceca.2012.10.005

[152] Colaluca IN, Tosoni D, Nuciforo P, Senic-Matuglia F, Galimberti V, Viale G, et al. NUMB controls p53 tumour suppressor activity. Nature. 2008;**451**:76-80. DOI: 10.1038/nature06412

[153] Elmore S. Apoptosis: A review of programmed cell death. Toxicologic Pathology. 2007;**35**:495-516. DOI: 10.1080/01926230701320337

[154] Berridge MJ, Bootman MD, Roderick HL. Calcium signalling: Dynamics, homeostasis and remodelling. Nature Reviews. Molecular Cell Biology. 2003;**4**:517-529. DOI: 10.1038/nrm1155

[155] Orrenius S, Zhivotovsky B, Nicotera P. Regulation of cell death: The calcium-apoptosis link. Nature Reviews. Molecular Cell Biology. 2003;**4**:552-565. DOI: 10.1038/nrm1150

[156] Pinton P, Giorgi C, Siviero R, Zecchini E, Rizzuto R. Calcium and apoptosis: ER-mitochondria Ca2+ transfer in the control of apoptosis. Oncogene. 2008;**27**:6407-6418. DOI: 10.1038/onc.2008.308

[157] Vanden Abeele F, Skryma R, Shuba Y, Van Coppenolle F, Slomianny C, et al. Bcl-2-dependent modulation of Ca(2+) homeostasis and store-operated channels in prostate cancer cells. Cancer Cell. 2002;**1**:169-179. DOI: 10.1016/s1535-6108(02)00034-x

[158] Prevarskaya N, Skryma R, Shuba Y. Ca2+ homeostasis in apoptotic resistance of prostate cancer cells. Biochemical and Biophysical Research Communications. 2004;**322**:1326-1335. DOI: 10.1016/j.bbrc.2004.08.037

[159] Vanoverberghe K, Vanden Abeele F, Mariot P, Lepage G, Roudbaraki M, et al. Ca2+ homeostasis and apoptotic resistance of neuroendocrine-differentiated prostate cancer cells. Cell Death and Differentiation. 2004;**11**:321-330. DOI: 10.1038/sj.cdd.4401375

[160] Kahl CR, Means AR. Regulation of cell cycle progression by calcium/calmodulin-dependent pathways. Endocrine Reviews. 2003;**24**:719-736. DOI: 10.1210/er.2003-0008

[161] Parekh AB, Putney JW Jr. Store-operated calcium channels. Physiological Reviews. 2005;**85**:757-810. DOI: 10.1152/physrev.00057.2003

[162] Wu TT, Peters AA, Tan PT, Roberts-Thomson SJ, Monteith GR. Consequences of activating the calcium-permeable ion channel TRPV1 in breast cancer cells with regulated TRPV1 expression. Cell Calcium. 2014;**56**:59-67. DOI: 10.1016/j.ceca.2014.04.006

[163] Nazıroğlu M, Çiğ B, Blum W, Vizlcr C, Buhala A, Marton A, et al. Targeting breast cancer cells by MRS1477, a positive allosteric modulator of TRPV1 channels. PLoS One. 2017;**12**:e0179950. DOI: 10.1371/journal.pone.0179950

[164] Raphael M, Lehen'kyi V, Vandenberghe M, Beck B,

Khalimonchyk S, Vanden Abeele F, et al. TRPV6 calcium channel translocates to the plasma membrane via Orai1-mediated mechanism and controls cancer cell survival. Proceedings of the National Academy of Sciences of the United States of America. 2014;**111**:E3870-E3879. DOI: 10.1073/pnas.1413409111

[165] Jain RK, Martin JD, Stylianopoulos T. The role of mechanical forces in tumor growth and therapy. Annual Review of Biomedical Engineering. 2014;**16**:321-346. DOI: 10.1146/annurev-bioeng-071813-105259

[166] Acerbi I, Cassereau L, Dean I, Shi Q, Au A, Park C, et al. Human breast cancer invasion and aggression correlates with ECM stiffening and immune cell infiltration. Integrative Biology. 2015;7:1120-1134. DOI: 10.1039/c5ib00040h

[167] Emon B, Bauer J, Jain Y, Jung B, Saif T. Biophysics of tumor microenvironment and cancer metastasis—A mini review. Computational and Structural Biotechnology Journal. 2018;**16**:279-287. DOI: 10.1016/j.csbj.2018.07.003

[168] Ramsey IS, Delling M, Clapham DE. An introduction to TRP channels. Annual Review of Physiology. 2006;**68**:619-647. DOI: 10.1146/annurev.physiol.68.040204.100431

[169] Saifeddine M, El-Daly M, Mihara K, Bunnett NW, McIntyre P, Altier C, et al. GPCR-mediated EGF receptor transactivation regulates TRPV4 action in the vasculature. British Journal of Pharmacology. 2015;**172**:2493-2506. DOI: 10.1111/bph.13072

[170] Adapala RK, Thoppil RJ, Ghosh K, Cappelli HC, Dudley AC, Paruchuri S, et al. Activation of mechanosensitive ion channel TRPV4 normalizes tumor vasculature and improves cancer therapy. Oncogene. 2016;**35**:314-322. DOI: 10.1038/onc.2015.83

[171] Sharma S, Goswami R, Zhang DX, Rahaman SO. TRPV 4 regulates matrix stiffness and TGF B1-induced epithelial-mesenchymal transition. Journal of Cellular and Molecular Medicine. 2019;**23**:761-774. DOI: 10.1111/jcmm.13972

[172] Sharma S, Goswami R, Rahaman SO. The TRPV4-TAZ mechanotransduction signaling axis in matrix stiffness- and TGFβ1-induced epithelial-mesenchymal transition. Cellular and Molecular Bioengineering. 2019;**12**:139-152. DOI: 10.1007/s12195-018-00565-w

[173] Cappelli HC, Kanugula AK, Adapala RK, Amin V, Sharma P, Midha P, et al. Mechanosensitive TRPV4 channels stabilize VE-cadherin junctions to regulate tumor vascular integrity and metastasis. Cancer Letters. 2019;**442**:15-20. DOI: 10.1016/j.canlet.2018.07.042

[174] Fiorio Pla A, Ong HL, Cheng KT, Brossa A, Bussolati B, Lockwich T, et al. TRPV4 mediates tumor-derived endothelial cell migration via arachidonic acid-activated actin remodeling. Oncogene. 2012;**31**:200-212. DOI: 10.1038/onc.2011.231

[175] Thoppil RJ, Cappelli HC, Adapala RK, Kanugula AK, Paruchuri S, Thodeti CK. TRPV4 channels regulate tumor angiogenesis via modulation of rho/rho kinase pathway. Oncotarget. 2016;7:25849-25861. DOI: 10.18632/oncotarget.8405

[176] Canales J, Morales D, Blanco C, Rivas J, Díaz N, Angelopoulo I, et al. TR(i)P to cell migration: New roles of TRP channels in mechanotransduction and cancer. Frontiers in Physiology.

2019;**10**:757. DOI: 10.3389/fphys.2019.00757

[177] Gogebakan B, Bayraktar R, Suner A, Balakan O, Ulasli M, Izmirli M, et al. Do fasudil and Y-27632 affect the level of transient receptor potential (TRP) gene expressions in breast cancer cell lines? Tumour Biology. 2014;**35**:8033-8041. DOI: 10.1007/s13277-014-1752-0

[178] Méndez-Samperio P. The human cathelicidin hCAP18/LL-37: A multifunctional peptide involved in mycobacterial infections. Peptides. 2010;**31**:1791-1798. DOI: 10.1016/j.peptides.2010.06.016

[179] Gambade A, Zreika S, Gueguinou M, Chourpa I, Fromont G, Bouchet AM, et al. Activation of TRPV2 and BKCa channels by the LL-37 enantiomers stimulates calcium entry and migration of cancer cells. Oncotarget. 2016;7:23785-23800. DOI: 10.18632/oncotarget.8122

[180] Habes C, Weber G, Goupille C. Sulfated glycoaminoglycans and proteoglycan syndecan-4 are involved in membrane fixation of LL-37 and its pro-migratory effect in breast cancer cells. Biomolecules. 2019;**9**:481. DOI: 10.3390/biom9090481

[181] Azimi I, Robitaille M, Armitage K, So CL, Milevskiy MJG, Northwood K, et al. Activation of the ion channel TRPV4 induces epithelial to mesenchymal transition in breast cancer cells. International Journal of Molecular Sciences. 2020;**21**:9417. DOI: 10.3390/ijms21249417

[182] Jiang Y, Gou H, Zhu J, Tian S, Yu L. Lidocaine inhibits the invasion and migration of TRPV6-expressing cancer cells by TRPV6 downregulation. Oncology Letters. 2016;**12**:1164-1170. DOI: 10.3892/ol.2016.4709

[183] Niiyama Y, Kawamata T, Yamamoto J, Omote K, Namiki A. Bone cancer increases transient receptor potential vanilloid subfamily 1 expression within distinct subpopulations of dorsal root ganglion neurons. Neuroscience. 2007;**148**:560-572. DOI: 10.1016/j.neuroscience.2007.05.049

[184] Li Y, Cai J, Han Y, Xiao X, Meng XL, Su L, et al. Enhanced function of TRPV1 via up-regulation by insulin-like growth factor-1 in a rat model of bone cancer pain. European Journal of Pain. 2014;**18**:774-784. DOI: 10.1002/j.1532-2149.2013.00420.x

[185] Tong Z, Luo W, Wang Y, Yang F, Han Y, Li H, et al. Tumor tissue-derived formaldehyde and acidic microenvironment synergistically induce bone cancer pain. PLoS One. 2010;**5**:e10234. DOI: 10.1371/journal.pone.0010234

[186] Gui Q, Xu C, Zhuang L, Xia S, Chen Y, Peng P, et al. A new rat model of bone cancer pain produced by rat breast cancer cells implantation of the shaft of femur at the third trochanter level. Cancer Biology & Therapy. 2013;**14**:193-199. DOI: 10.4161/cbt.23291

[187] Austin M, Elliott L, Nicolaou N, Grabowska A, Hulse RP. Breast cancer induced nociceptor aberrant growth and collateral sensory axonal branching. Oncotarget. 2017;**8**:76606-76621. DOI: 10.18632/oncotarget.20609

[188] Ghilardi JR, Rohrich H, Lindsay TH, Sevcik MA, Schwei MJ, Kubota K, et al. Selective blockade of the capsaicin receptor TRPV1 attenuates bone cancer pain. The Journal of Neuroscience. 2005;**25**:3126-3131. DOI: 10.1523/JNEUROSCI.3815-04.2005

[189] Xu Q, Zhang XM, Duan KZ, Gu XY, Han M, Liu BL, et al. Peripheral

TGF-β1 signaling is a critical event in bone cancer-induced hyperalgesia in rodents. The Journal of Neuroscience. 2013;**33**:19099-19111. DOI: 10.1523/JNEUROSCI.4852-12.2013

[190] Pan HL, Zhang YQ, Zhao ZQ. Involvement of lysophosphatidic acid in bone cancer pain by potentiation of TRPV1 via PKCε pathway in dorsal root ganglion neurons. Molecular Pain. 2010;**6**:85. DOI: 10.1186/1744-8069-6-85

[191] Gautier M, Dhennin-Duthille I, Ay AS, Rybarczyk P, Korichneva I, Ouadid-Ahidouch H. New insights into pharmacological tools to TR(i)P cancer up. British Journal of Pharmacology. 2014;**171**:2582-2592. DOI: 10.1111/bph.12561

[192] Carnevale V, Rohacs T. TRPV1: A target for rational drug design. Pharmaceuticals (Basel, Switzerland). 2016;**9**(3):52. DOI: 10.3390/ph9030052

[193] Fu S, Hirte H, Welch S, Ilenchuk TT, Lutes T, Rice C, et al. First-in-human phase I study of SOR-C13, a TRPV6 calcium channel inhibitor, in patients with advanced solid tumors. Investigational New Drugs. 2017;**35**:324-333. DOI: 10.1007/s10637-017-0438-z

[194] Koivisto AP, Belvisi MG, Gaudet R, Szallasi A. Advances in TRP channel drug discovery: From target validation to clinical studies. Nature Reviews. Drug Discovery. 2022;**21**:41-59. DOI: 10.1038/s41573-021-00268-4

[195] Saldías MP, Maureira D, Orellana-Serradell O, et al. TRP channels interactome as a novel therapeutic target in breast Cancer. Frontiers in Oncology. 2021;**11**:621614. DOI: 10.3389/fonc.2021.621614

[196] Díaz-Laviada I, Rodríguez-Henche N. The potential antitumor effects of capsaicin. Progress in Drug Research. 2014;**68**:181-208. DOI: 10.1007/978-3-0348-0828-6_8

[197] Pecze L, Jósvay K, Blum W, Petrovics G, Vizler C, Oláh Z, et al. Activation of endogenous TRPV1 fails to induce overstimulation-based cytotoxicity in breast and prostate cancer cells but not in pain-sensing neurons. Biochimica et Biophysica Acta. 2016;**1863**:2054-2064. DOI: 10.1016/j.bbamcr.2016.05.007

[198] Ferreira AK, Tavares MT, Pasqualoto KF, de Azevedo RA, Teixeira SF, Ferreira-Junior WA, et al. RPF151, a novel capsaicin-like analogue: In vitro studies and in vivo preclinical antitumor evaluation in a breast cancer model. Tumour Biology. 2015;**36**:7251-7267. DOI: 10.1007/s13277-015-3441-z

[199] Erin N. Role of sensory neurons, neuroimmune pathways, and transient receptor potential vanilloid 1 (TRPV1) channels in a murine model of breast cancer metastasis. Cancer Immunology, Immunotherapy. 2020;**69**:307-314. DOI: 10.1007/s00262-019-02463-0

[200] De La Chapa J, Valdez M, Ruiz F 3rd, Gonzales K, Mitchell W, McHardy SF, et al. Synthesis and SAR of novel capsazepine analogs with significant anti-cancer effects in multiple cancer types. Bioorganic & Medicinal Chemistry. 2019;**27**:208-215. DOI: 10.1016/j.bmc.2018.11.040

[201] Ligresti A, Moriello AS, Starowicz K, Matias I, Pisanti S, De Petrocellis L, et al. Antitumor activity of plant cannabinoids with emphasis on the effect of cannabidiol on human breast carcinoma. The Journal of Pharmacology and Experimental Therapeutics. 2006;**318**:1375-1387. DOI: 10.1124/jpet.106.105247

[202] Farsandaj N, Ghahremani MH, Ostad SN. Role of cannabinoid

and vanilloid receptors in invasion of human breast carcinoma cells. Journal of Environmental Pathology, Toxicology and Oncology. 2012;**31**:377-387. DOI: 10.1615/jenvironpatholtoxicoloncol.2013005859

[203] Sakallı Çetin E, Nazıroğlu M, Çiğ B, Övey İS, Aslan KP. Selenium potentiates the anticancer effect of cisplatin against oxidative stress and calcium ion signaling-induced intracellular toxicity in MCF-7 breast cancer cells: Involvement of the TRPV1 channel. Journal of Receptor and Signal Transduction Research. 2017;**37**:84-93. DOI: 10.3109/10799893.2016.1160931

[204] Nur G, Nazıroğlu M, Deveci HA. Synergic prooxidant, apoptotic and TRPV1 channel activator effects of alpha-lipoic acid and cisplatin in MCF-7 breast cancer cells. Journal of Receptor and **Signal** Transduction Research. 2017;**37**:569-577. DOI: 10.1080/10799893.2017.1369121

[205] Deveci HA, Nazıroğlu M, Nur G. 5-fluorouracil-induced mitochondrial oxidative cytotoxicity and apoptosis are increased in MCF-7 human breast cancer cells by TRPV1 channel activation but not hypericum perforatum treatment. Molecular and Cellular Biochemistry. 2018;**439**:189-198. DOI: 10.1007/s11010-017-3147-1

[206] Heilborn JD, Nilsson MF, Jimenez CI, Sandstedt B, Borregaard N, Tham E, et al. Antimicrobial protein hCAP18/LL-37 is highly expressed in breast cancer and is a putative growth factor for epithelial cells. International Journal of Cancer. 2005;**114**:713-719. DOI: 10.1002/ijc.20795

[207] Darby WG, Grace MS, Baratchi S, McIntyre P. Modulation of TRPV4 by diverse mechanisms. The International Journal of Biochemistry & Cell Biology. 2016;**78**:217-228. DOI: 10.1016/j.biocel.2016.07.012

[208] Grace MS, Bonvini SJ, Belvisi MG, McIntyre P. Modulation of the TRPV4 ion channel as a therapeutic target for disease. Pharmacology & Therapeutics. 2017;**177**:9-22. DOI: 10.1016/j.pharmthera.2017.02.019

[209] Landowski CP, Bolanz KA, Suzuki Y, Hediger MA. Chemical inhibitors of the calcium entry channel TRPV6. Pharmaceutical Research. 2011;**28**:322-330. DOI: 10.1007/s11095-010-0249-9

[210] Bowen CV, DeBay D, Ewart HS, Gallant P, Gormley S, Ilenchuk TT, et al. In vivo detection of human TRPV6-rich tumors with anti-cancer peptides derived from soricidin. PLoS One. 2013;**8**:e58866. DOI: 10.1371/journal.pone.0058866

[211] Soricimed. Soricimed Completes Phase 1 Trial of SOR-C13 in Advanced Solid Tumour Cancers. 2015. Available from: http://www.soricimed.com [Accessed: August 2016]

[212] Stewart JM. TRPV6 as a target for cancer therapy. Journal of Cancer. 2020;**11**:374-387. DOI: 10.7150/jca.31640

[213] Santoni G, Maggi F, Morelli MB, Santoni M, Marinelli O. Transient receptor potential cation channels in cancer therapy. Medical Sciences (Basel). 2019;7:12. DOI: 10.3390/medsci7120108

[214] Steinritz D, Stenger B, Dietrich A, Gudermann T, Popp T. Trps in Tox: Involvement of transient receptor potential-channels in chemical-induced organ toxicity—a structured review. Cell. 2018;7:235-256. DOI: 10.3390/cells7080098

Chapter 5

SK Channels and Heart Disease

Katherine Zhong, Shawn Kant, Frank Sellke and Jun Feng

Abstract

Extensive evidence indicates that small-conductance Ca^{2+}-activated K^{+} channels (SK channels) help regulate cardiac rhythm and myocardial function in physiological and pathophysiological conditions. This chapter will begin by discussing the basic physiology of SK channel expression, localization, and activation under normal conditions, before proceeding to address the impact of SK channel dysfunction on a variety of cardiac pathologies including atrial fibrillation (AF), ventricular arrhythmias (VA), cardiac hypertrophy/heart failure (HF) and myocardial ischemia/reperfusion (IR) injury. The critical role of aberrant SK channel regulation will also be discussed to establish unifying mechanisms of SK channel pathology across these different conditions. Several animal model and human tissue experiments suggest that pharmacologic modulation of SK channel function may be beneficial in controlling AF, VA, cardiomyopathy and myocardial IR injury. Therefore, targeting SK channels may represent a promising new therapeutic avenue for treating a variety of cardiovascular disease states.

Keywords: SK channel, small-conductance Ca^{2+}-activated K^{+} channels, cardiac rhythm, cardiac function, atrial fibrillation, ventricular arrhythmias, cardiac hypertrophy, heart failure, ischemia reperfusion injury

1. Introduction

The coordinated activity of cell membrane ion channels forms the basis of the cardiomyocyte action potential sequence, driving myocardial contractility and resulting in cardiac output. Each stage of the myocardial action potential relies on opening and closing of specific groups of ion channels to propagate excitatory stimuli through the heart [1–4]. Most myocardial action potentials begin with transmission of spontaneous impulses generated by pacemaker cells (e.g., the sinoatrial node [SAN] and atrioventricular node [AVN]) to atrial and ventricular myocytes that are at resting membrane potential (phase 4, around −85 to −90 mV). Sinus node impulses reach atrial myocytes via intercellular gap junctions, leading to depolarization to threshold. Myocytes then transition to phase 0, a fast upstroke driven by opening of voltage-gated sodium channels that facilitate a strong inward sodium current that rapidly depolarizes the cardiomyocyte to a membrane potential of over 20 mV, at which point sodium channels close and transient outward potassium channels open, producing a brief repolarizing current (phase 1).

However, the strongly depolarized cardiomyocyte membrane also triggers opening of L-type voltage gated calcium channels (LTCC) that facilitate influx of calcium.

Calcium influx balances potassium efflux through transient outward and delayed-rectifier potassium channels (which also open during this period), leading to the phase 2 plateau of the cardiac action potential. It is during phase 2 that calcium entering cardiomyocytes triggers calcium-induced calcium release from sarcoplasmic reticulum (SR) stores via calcium-induced calcium release, mediated by calcium interaction with ryanodine receptor (RyR) calcium channels located on the SR [2]. Increasing levels of intracellular calcium relieves troponin-tropomyosin inhibition of cardiomyocyte actin-myosin cross bridge formation, leading to cross-bridge cycling and myocardial contraction. Finally, cardiomyocytes transition to phase 3 repolarization, where calcium channels close and delayed rectifier potassium channels remain open, allowing membrane potential to return to its resting state (phase 4) and terminating the action potential.

The repolarization phase of the myocardial action potential has received extensive attention in biomedical research because aberrations in this phase are intimately related to disruptions in myocardial excitability and overall control of synchronous, regulated myocardial contraction. Until recently, phase 3 repolarization was thought to be almost entirely driven by delayed rectifier potassium channels that opened alongside calcium channels during phase 2 and remained open following calcium channel closure, subsequently driving membrane potential towards the potassium equilibrium potential.

However, a growing body of evidence suggests that another type of potassium channel, the small-conductance Ca^{2+}-activated K^+ channels (SK channel), might also have an important role in myocardial repolarization [5–8]. SK channels are cell membrane potassium channels that are exquisitely sensitive to calcium, opening in response to elevated intracellular calcium and facilitating an outward potassium current. Several studies have shown that SK channels are active during the late repolarization phase of the cardiac action potential, likely assisting delayed rectifier potassium channels by amplifying the outward potassium current [5, 6]. On their own, SK channels may represent a form of feedback control of excess myocardial excitability, given their close responsiveness to calcium—which governs cardiomyocyte contraction.

Conversely, SK channel dysfunction may contribute to cardiovascular pathology in a manner reminiscent of how dysfunction of other major myocardial ion channels (especially the delayed rectifier potassium channel) is implicated in cardiovascular disease. Indeed, the past two decades have given rise to an abundance of research attempting to characterize the role of SK channels in numerous cardiovascular disease states and explore whether modulation of SK channels may be a potential therapeutic tool that deserves clinical attention. Hence the focus of this chapter, which will discuss the history and basic physiology of myocardial SK channels before delving into the current literature concerning SK channel dysfunction in four major areas: atrial fibrillation (AF), ventricular tachyarrhythmias (VA), heart failure (HF), and ischemia-reperfusion (IR) injury.

2. Identification of SK channels in the heart

Small-conductance Ca^{2+}-activated K^+ channels (SK channels), encoded by KCNN genes, are a family of K^+ channels that have a small single channel conductance (10–20 pS with symmetrical solutions), and are gated by intracellular Ca^{2+} concentrations [9]. SK channels were identified using apamin, a bee venom toxin that selectively blocks SK channels [9]. SK channels were first identified in rat and human brain tissue [10]. Since their discovery, SK channels have been identified in a variety of tissues including the nervous system, blood, epithelial cells, skeletal muscle and endothelial (vessel) cells [11, 12]. SK channels are classified into three isoforms: SK1, SK2 and SK3, based on their gene origins and different sensitivities to apamin. The SK1 channel is encoded by the KCNN1 gene, located on chromosome 19, and is moderately sensitive to apamin. The SK2 channel is encoded by the KCNN2 gene located on chromosome 5 and has the strongest affinity for apamin. The SK3 channel is encoded by the KCNN3 gene located on chromosome 1 and is moderately sensitive to apamin [8, 13]. The intermediate-conductance Ca^{2+}-activated K^+ channel (IK or SK4) is similar to SK channels and is encoded by the KCNN4 gene. SK4 has a slightly larger single conductance (12–42 pS) and a higher affinity for intracellular Ca^{2+} than other SK channels [8, 13].

SK channels were first identified in the heart in 2003 [14, 15], and their distribution and functions in heart tissue have since been extensively studied [16–18]. SK channels have been found in both atrial and ventricular tissues in animals and humans [7, 12, 19–33]. In mouse atrial and ventricular myocytes, quantification of SK1 and SK3 transcripts showed a higher level of SK1 expression in atria versus ventricles, while SK3 is expressed at a simar level in atria and ventricles [12]. Apamin-sensitive Ca^{2+}-activated K^+ currents (SK currents) have been detected in the pulmonary veins (PV), Bachmann's bundle (BB), sinoatrial node (SAN) myocytes, and mouse atrioventricular nodes (AVN) [34–36]. SK channels have also been found expressed intracellularly in the ventricular mitochondrial membranes of guinea pigs, rats and humans [37, 38]. Indeed, transcripts of both SK2 and SK3 isoforms are found in the inner mitochondrial membrane (IMM), but not SK1 [38]. Notably, although there are fewer SK channels in the ventricles, under certain pathological conditions such as HF and chronic myocardial infarction (MI), SK currents in ventricles are upregulated [39, 40]. The results are summarized in **Table 1**.

Species	Tissue/cardiomyocyte	mRNA	Protein	Ion channel recording	SK channels	Reference
Mouse	Atrial and ventricular myocytes	Higher level of expression of SK1 in atria versus ventricles; SK3 is expressed at a simar level in both	Protein expression of SK1 and low level of SK3	SK current is more prominent in atria than in ventricles	SK1, SK3	[12]
	Anatomical atrioventricular nodes (AVN)	N/A	SK2 channel protein detected in AVN cells	SK current recorded in AVN cells	SK2	[36]
	SAN cells	Presence of the transcripts of SK1, SK2 and SK3	All three isoforms are present and are preferentially distributed along the Z line, especially SK3	SK currents recorded	SK1, SK2, SK3	[41]

Species	Tissue/ cardiomyocyte	mRNA	Protein	Ion channel recording	SK channels	Reference
Rabbit	Pulmonary vein (PV) and Bachmann's bundle (BB)	SK2 and SK3 mRNA detected in both PV and BB	Presence of SK2 and SK3 protein in both PV and BB	SK current recorded in both PV and BB	SK2, SK3	[6]
	Ventricles	N/A	N/A	SK currents are not significant in normal ventricles but are upregulated in failing ventricles	N/A	[39]
	Pulmonary vein (PV) and sinoatrial node (SAN) myocyte	N/A	N/A	SAN exhibits greater SK currents than PV	N/A	[35]
	Ventricles	N/A	SK2 are more abundantly present in the Purkinje cells (PCs) than in the ventricular myocytes	Apamin prolonged APD in PCs. SK current density was larger in PCs than in ventricular myocytes	SK2	[42]
Guinea pig	Ventricular mitochondria	N/A	SK2 and SK3 channel proteins were present in IMM	SK current recorded	SK2, SK3	[16]
Human	Right and left atrial appendages	mRNAs of SK1, SK2 and SK3 were detected	SK1, SK2 and SK3 were expressed in human atrial myocytes	SK current detected	SK1, SK2, SK3	[27]
	Atrial and ventricular tissue	SK1 had a higher level of expression in atria compared to ventricle, while the expression level of SK2 and SK3 were not different in atria vs. ventricle	N/A	N/A	SK1, SK2, SK3	[25]
	Atrial tissues	N/A	SK3 polypeptide level was unchanged pre- and post-cardioplegic arrest	N/A	SK3	[19]
	Right atrial tissues	N/A	Insignificant decrease in SK3	Diabetes significantly	SK3	[20]

Species	Tissue/ cardiomyocyte	mRNA	Protein	Ion channel recording	SK channels	Reference
			protein level in diabetic myocardium vs. nondiabetics	reduced SK currents		
	Right atrial tissues	N/A	Neither cardioplegic arrest nor diabetes resulted in a significant change of SK3 protein level	N/A	SK3	[21]
	Human atrial tissue, coronary arterioles and coronary artery endothelial cells	No significant difference in SK3 mRNA levels in diabetic atrial tissues and endothelial cells vs. nondiabetics	No significant difference in SK3 protein levels in diabetic atrial tissues and endothelial cells vs. nondiabetics	N/A	SK3	[33]
Multiple species	Mouse atrial and ventricular and human atrial myocytes	Presence of SK2 mRNA in human and mouse cardiac myocytes	Presence of SK2 channel	SK current was more prominent in atria versus ventricles	SK2	[14]
	Guinea pig, rat and human ventricular mitochondria	mRNAs of SK2 and SK3, but not SK1, are present in guinea pig ventricular myocytes. SK3 mRNAs are present in human ventricular tissues.	SK3 is expressed in human and guinea pig ventricular IMM	N/A	SK3	[38]

Table 1.
Functional expression of SK channels in the heart.

3. Physiological functions of SK channels in the heart

In neuronal cells, SK channels contribute to the slow afterhyperpolarization following an action potential [43]. The physiologic functions of SK channels in the heart were not clear until Xu et al. first reported their role in the repolarization of cardiac myocytes. Apamin-treated cardiac cells show significantly longer action potential durations (APD) due to slower repolarization compared to control groups, indicating SK channel involvement in cardiac repolarization. These effects are more prominent in atrial versus ventricular cells, likely due to the higher level of expression of SK

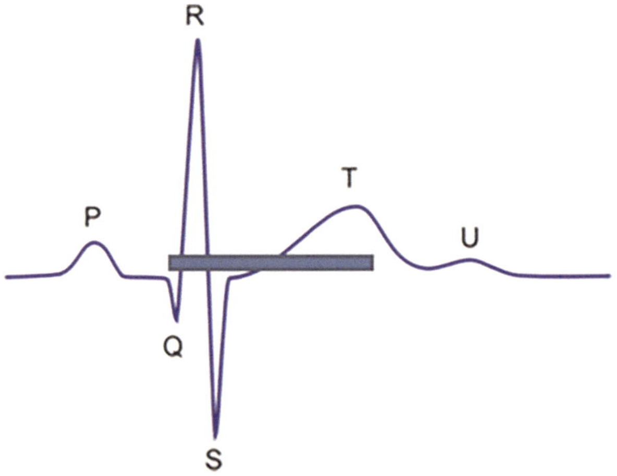

Figure 1.
Cardiac action potential. P = atrial depolarization; QRS = ventricular depolarization; T = ventricular repolarization. The significance of the U wave is largely unknown, although may represent Purkinje fiber repolarization. The QT interval indicated by the dark bar between "Q" and "T". SK channels have been shown to influence the duration of the QT interval of cardiomyocyte action potentials. Diagram courtesy of ECGpedia.org (https://en.ecgpedia.org/wiki/Action_potential) [45].

channels in atria [14]. In another study, the AVN in SK2-null mutant mice show decreased firing frequency and longer APD compared with controls, while mice overexpressing SK2 channels show shorter APD [36]. Apamin application results in decreased action potential firing frequency akin to that seen in SK2-null mutants [36].

Consistently, overexpression of SK3 channels also results in significantly shorter APD, suggesting a similar role for different isoforms of SK channels in the repolarization process in cardiac cells [44]. The prolongation of APD resulting from apamin has also been recorded in human right and left atrial appendages, rabbit PV and SAN, and mouse SAN cells [27, 35, 41]. These studies consistently show that SK channels contribute to repolarization and shorten the action potentials in normal hearts, especially in the atria where they are more densely expressed than in the ventricles. In SAN, SK blockade also leads to significant depolarization of the maximal diastolic potential (MDP) and a decrease in the diastolic depolarization slope [41]. Notably, in PVs with intact endothelium, SK channels contribute to hyperpolarization and vessel dilation [35]. **Figure 1** shows the cardiac action potential including the QT interval affected by SK channels.

Meanwhile, with respect to mitochondria, DCBE, an SK channel opener, reduces mitochondrial injury when given before cardiac ischemia, suggesting that SK channel opening may protect the heart and mitochondria against ischemia-reperfusion (IR) injury. SK channels have also been found to reduce oxidative stress by reducing mitochondrial Ca^{2+} overload [38].

4. Signaling pathway

The Ca^{2+} that activates SK channels comes from various intracellular and extracellular sources. In hippocampal neurons, SK channels are coupled to the LTCC using microdomains of submembrane calcium, and Ca^{2+} influx through LTCC activates SK channels [46]. Similar co-localization of LTCC and SK channels can be found in cardiac myocytes, where SK2 channels couple with LTCCs through α-actinin2. This suggests that, in cardiac myocytes, SK2 channels are activated by local subsarcolemmal Ca^{2+} that enters the cell through LTCCs [47]. Besides LTCCs, the SR also plays a role in providing the Ca^{2+} needed for activating SK channels. In mouse cardiac myocytes, the type 2 ryanodine receptor (RyR2) mediates intracellular Ca^{2+} release from the SR, which activates SK channels [48]. It is possible that in cardiac myocytes, SK channels could be activated by either Ca^{2+} influx through voltage-gated Ca^{2+} channels or through release of intracellular Ca^{2+} from SR stores.

SK channels are gated by intracellular Ca^{2+} through interactions between the channel α-subunits and calmodulin (CaM) [49]. The intracellular C-terminal domain of the SK channels immediately adjacent to the sixth transmembrane segment (S6) consists of four α-helices, named A, B, C and D, that are critical to the channel gating. CaM binds to channel regions A-D constitutively, whereas CaM binding to regions B-C and B-D is Ca^{2+} dependent. This indicates that regions B-C and B-D are involved in the Ca^{2+} gating mechanism [49]. Ca^{2+} binds to the EF hands in the N-lobe of CaM, which leads to conformational changes and opening of SK channels [50].

Besides Ca^{2+}-activated channel gating, CaM is also critical in the regulation of cell surface expression of the SK channels, independent of the binding of Ca^{2+} [51]. Several cytoskeletal proteins, including α-actinin2, filamin A, and MLC2, are important in SK2 channel trafficking [52–54]. Notably, cell membrane localization of SK2 channels is Ca^{2+}-dependent when the channels are co-expressed with α-actinin2. An increase in intracellular Ca^{2+} such as in AF is predicted to increase the expression of SK2 channels and lead to shortened APD and maintenance of the arrhythmias [54].

Casein kinase 2 (CK2) and protein phosphatase 2A (PP2A) are critical components of the SK channels that regulate the Ca^{2+} sensitivity of the channels by phosphorylating or dephosphorylating CaM. CK2 decreases the Ca^{2+} sensitivity of closed SK channels, while PP2A increases the Ca^{2+} sensitivity of open SK channels [9]. Increased expression of PP2A and decreased co-localization of CK2 with SK2 may be the underlying mechanism of increased sensitivity of apamin-sensitive K^+ current to intracellular Ca^{2+} in HF, as shown in a volume-overload HF rat model [55].

Various mechanisms of SK channel regulation may have clinical significance with respect to SK channel pathology seen in certain disease states. For example, upregulation of SK channels in ventricular myocytes in cardiac hypertrophy has been shown to result from phosphorylation by both calcium/calmodulin-dependent protein kinase II (CaMKII) and protein kinase A (PKA) [28, 56]. In addition, microRNA (miRNA) also plays a role in cardiac SK channels regulation. MicroRNA 499 (miR-499) is upregulated in atrial tissue from patients with permanent AF, resulting in increased binding to the KCNN3 gene and downregulation of SK2 channel expression [26].

5. SK channels in heart disease

5.1 Atrial fibrillation

5.1.1 Overview

Atrial fibrillation (AF), a condition characterized by rapid and disorganized atrial activation, is the most common cardiac arrhythmia with a prevalence of 1–2% among the general population [57]. A variety of risk factors have been linked to the development of AF, such as age, male sex, obesity, hypertension, heart failure, structural heart disease (valvulopathy, CHF, MI), diabetes, hyperthyroidism, and family history [58]. Of these factors, age and sex carry the highest risk: males have a 1.5–2× greater chance of developing AF than females, and individuals between ages 40–55 carry a lifetime risk of 22–26% [59].

AF is classified into several categories based on the duration of the AF episode(s). These include paroxysmal AF, persistent AF, long-standing persistent AF, and permanent AF [60]. The term "lone AF" remains a diagnosis of exclusion and has been used to describe AF in younger patients without a prior history of structural heart disease or cardiovascular risk factors [61]. The specific molecular and electrophysiological mechanisms that underly AF are highly complex and remain poorly understood. Nonetheless, most current conceptual frameworks of the pathogenesis of AF involve a combination of structural remodeling, electrical remodeling, and autonomic remodeling that generate atrial reentry circuits, rotors (localized electrical spiral waves), and ectopic impulse generation [57].

Regarding electrical remodeling, a variety of altered ionic currents can be seen in AF, including increased activity of LTCC, increased activity of inward rectifier potassium channels (KiR), and decreased function of gap junctions [62–65]. Excessive LTCC activity may trigger excessive cardiomyocyte SR Ca^{2+} release via RyR activation, leading to hyperexcitability. Excessive KiR activity may alter atrial myocyte resting potential and phase 3 activation, resulting in reduced atrial refractoriness. Gap junction defects may slow atrial conduction velocity, which when combined with atrial myocyte hyperexcitability and decreased refractoriness favors reentry, ectopic foci, and initiation of AF.

Structural remodeling in AF largely involves fibrosis and atrial dilation. Atrial fibrosis may be due to several factors such as aging, myocardial infarction, volume overload, or aberrant renin-angiotensin-aldosterone system (RAAS) activity [66–70]. Proliferation of fibroblasts and extracellular matrix in fibrotic atria may create barriers to electrical conduction that interfere with cardiomyocyte electrical coupling, creating conduction abnormalities and variable action potential durations that predispose to ectopic activity and reentrant circuits [71, 72]. In addition, larger atria have increased odds of developing reentrant circuits [73]. Finally, autonomic remodeling in AF involves increased sympathetic or parasympathetic tone in the atria that influences atrial tachypacing [74–77].

5.1.2 SK channels and AF heritability

The heritability of AF has been widely studied over the past two decades, and lone AF appears to have a greater heritable component than structural AF [78]. Recent genome-wide association studies (GWAS) have identified several single nucleotide polymorphisms (SNPs) that are associated with increased risk of AF. Some examples include rs2200733 on chromosome 4q25, rs2106261 on chromosome 16q22, rs10824026 on chromosome 10q22, and rs7394190 on chromosome 10q22 [61, 78, 79]. Overall, GWAS studies of AF atrial tissue have found mutations in genes coding for ion channels, gap junction connexin proteins, nuclear membrane components, calcium homeostasis, and cardio-genesis [80].

Potassium channel gene mutations are of special interest in AF because of the important role of potassium channels in maintaining the resting membrane potential and facilitating repolarization after generation of the action potential. For example, gain of function mutations have been found in KCNQ1 (the alpha subunit of the inward rectifier cardiac potassium channel Kv7.1), KCNH2 (voltage gated potassium channel Kv11.1), KCND3 (alpha subunit of the voltage-gated potassium channel Kv4.3), KCNJ2 (inward rectifying potassium channel Kir2.1), and KCNA5 (voltage-gated potassium channel Kv1.5) [81–85]. All of these changes lead to increased inward potassium currents in atrial myocytes, which may shorten APD, QT intervals, and the effective refractory periods. In 2010, Ellinor et al. in their GWAS discovered a novel AF susceptibility locus on chromosome 1q21 among Caucasian individuals. The most significant SNP at this locus, rs13376333, can be found in the intron between the first and second exon of KCNN3, which encodes the SK3 channel [61]. Rs13376333 is strongly associated with lone AF, bearing an odds ratio of 1.56 ($P = 6.3 \times 10^{-12}$). The association of rs13376333 with more typical forms of AF is also significant, albeit weaker (OR = 1.13, P = 0.006) [61].

Similar results have been reported by several other labs using genomic data from identical or different ethnic groups. Indeed, Chang et al. report significant associations between rs13376333 and risk of lone (OR 3.02) and structural (OR 2.18) AF among a Taiwanese cohort, with even stronger odds ratios than those reported by Ellinor et al. [86]. Among a Han Chinese cohort, Luo et al. found that the frequency of rs13376333 at KCNN3 was significantly higher in lone AF than in controls, although there were no significant differences in rs13376333 frequency between total AF patients and controls [87]. Curiously, an earlier study by Li et al. failed to produce a significant association between rs13376333 and AF among Han Chinese AF individuals [88]. It is possible that different ethnic demographic groups may exhibit different risk propensities with respect to SNPs, which would imply contribution of some form of gene-environment interplay that either diminishes or enhances genetic effects. Given the complexity of AF as a disease process, this is a very likely scenario and requires further investigations involving larger sample sizes and different demographic groups. Besides rs13376333, another SNP at KCNN3, rs1131820, has also been associated with increased risk of lone AF in individuals of Danish ethnicity, with an OR of 2.85 [89].

In addition to KCNN3, a weak association of AF with KCNN2 has also been discovered in a Han Chinese cohort, at the SNP rs13184658 [90]. How variations in KCNN alleles specifically affect the risk of AF remains unclear. A recent study revealed that rs13376333 is associated with increased mRNA expression of KCNN3 in human atrial tissue [91]. In addition, given that most other potassium channel mutations observed in AF are gain of function mutations, perhaps the SK channel SNPs follow the same trend. Further study of the KCNN3 locus could help reveal pathways underlying the association between KCNN3 and AF, with the potential of developing novel treatments of AF that target KCNN3.

5.1.3 SK expression and electrophysiology in AF

At the mRNA and protein levels, there is mixed evidence about whether SK channels are over or under-expressed in atrial remodeling in human and animal models of AF. Ozgen et al. first reported that SK channels are associated with initiation of atrial remodeling. Using a burst-paced rabbit atrium, they showed that SK2 mRNA and protein levels were upregulated in the region of the left atrium where the pulmonary veins empty; this suggests that SK channels have a role in burst-pacing induced APD shortening [34]. Likewise, Qi et al. reported upregulation of SK1 and SK2 protein induced by atrial tachypacing in dog PVs and left atrial cells; SK2 mRNA expression was also increased, although no significant changes in SK1 mRNA were observed, suggesting that overexpression of SK1 may be the result of posttranslational modifications or altered membrane trafficking of SK1 channels [23]. In dopamine tautomerase-deficient mice induced to exhibit AF by apamin administration, Tsai et al. observed increased protein and mRNA levels of SK1 and SK3 in mouse right atrial tissue [92].

These results seem to contradict other studies that reported decreased expression of SK channels in human and animal AF. Darkow et al. found decreased expression of KCNN2 (SK2) mRNA in atrial tissue of patients with AF when compared with healthy controls [25]. Similarly, transcripts of KCNN1–3 were downregulated in paroxysmal AF and chronic AF patients at a similar level [93]. Rahm et al. also examined KCNN mRNA expression in a pig model of atrial tachypacing-induced AF with reduced left ventricular function and found decreased expression of KCNN2 and KCNN3 with normal levels of KCNN1 [93]. Furthermore, Fan et al. report decreased mRNA and protein levels of SK1, SK2, and SK3 in atrial appendage tissue from humans with chronic AF [32]. Another study of atrial tissue from chronic AF patients by Yu et al. replicated these results with respect to SK1 and SK2 mRNA and protein expression, although no differences were observed in SK3 expression at either level [27]. Finally, Skisbye et al. observed reduced SK2 and SK3 expression in chronic AF human atrial tissue, with no changes in SK1 expression [7]. The research findings of the up/down regulation of SK channels in AF were summarized in **Table 2**.

Ozgen et al. first discovered that burst pacing induced increased SK2 channel trafficking to the cell membrane and increased SK currents in rabbit pulmonary veins [34]. In their dog model, Qi et al. found that atrial tachypacing enhanced SK currents and single-channel open probabilities [23]. In contrast, Yu et al. reported decreased SK currents in right and left atrial appendage tissue from patients with chronic AF, alongside decreased SK1 and SK2 mRNA and protein levels (discussed earlier) [27]. Another possible mechanism of increased SK currents in AF is enhanced activity of

Species	Model	Up/down regulation of SK channels	Reference
Rabbit	Burst-paced rabbit atrium	SK2 mRNA and protein levels were upregulated in the left atrium	[34]
Dog	Atrial tachypacing in dog pulmonary veins and left atrial cells	Upregulation of SK1 and SK2 protein; upregulated SK2 mRNA expression	[23]
Mouse	Induced AF in dopamine tautomerase-deficient mice	Increased protein and mRNA levels of SK1 and SK3 in right atrial tissue	[92]
Pig	Atrial tachypacing-induced AF with reduced left ventricular function	Decreased expression of KCNN2 and KCNN3 with normal levels of KCNN1	[93]
Human	Patients with AF	Decreased expression of SK2 mRNA in atrial tissue	[25]
	Paroxysmal AF and chronic AF patients	Transcripts of KCNN1–3 were downregulated	[93]
	Atrial appendage tissue from chronic AF patients	Decreased mRNA and protein levels of SK1, SK2, and SK3	[32]
	Atrial tissue from chronic AF patients	Decreased mRNA and protein levels of SK1, SK2; no difference in SK3 expression levels	[27]
	Atrial tissue from chronic AF patients	Reduced SK2 and SK3 expression; no changes in SK1 expression	[7]

Table 2.
Up/down regulation of SK channels in atrial fibrillation (AF).

CaMKII, which is required for SK channel calcium-dependent activation. CaMKII exhibited significantly increased expression alongside, increased intracellular calcium levels in human AF tissue studied by Fan et al. [32].

Approached from a different angle, a mouse model of SK3 overexpression also showed considerable shortening of APD and increased SK channel currents in atrial myocytes [44]. Likewise, mice engineered to overexpress SK2 channels displayed significant atrioventricular nodal dysfunction, manifesting as increased firing frequency, and shortening of spontaneous action potential, while SK2 ablation eliminated these effects [36]. Furthermore, SK2 knockout mouse models studied by Li et al. exhibited the opposite effects: significant prolongation of atrial myocyte APD among homozygous and heterozygous mutants, with homozygous mutants having an increased susceptibility to AF [94].

Inconsistencies regarding observed SK channel expression and activity in AF across different studies might be due to a variety of factors including the specific method of AF induction, differences in atrial tissue characteristics among different species of animal models, patient population demographic differences, and different stages or durations of AF. It is also possible that SK channels are initially upregulated in AF before being downregulated due to atrial remodeling. There are several reported mechanisms by which SK channels are regulated in AF, including histone deacetylase related epigenetic mechanisms [95], miRNA [26], and CaMKII [32].

5.1.4 Pharmacologic modulation of SK channels in AF

Pharmacologic studies of SK channel modulation provide additional insights into the role of SK channels in AF, although the results are complicated. First, the IK antagonist HMR1556 preserved hemodynamic stability in pigs induced by atrial burst pacing to express persistent AF [96]. At time of sacrifice, HMR1556 treated pigs also exhibited significantly higher left ventricular ejection fraction than untreated pigs, along with significantly longer right atrial APD [96]. Next, application of the SK channel inhibitor apamin increased spontaneous action potential generation in isolated rabbit PVs while decreasing spontaneous activity and prolonging APD in sinoatrial nodal myocytes [35]. However, in isolated canine left atrial tissue, apamin treatment significantly increased APD heterogeneity and proved to be pro-arrythmogenic [24].

The SK channel inhibitor NS8593 increased APD and effective refractory in right atrial appendage tissue of patients with AF [7]. Identical results were found by Qi et al. in a dog model of AF [23]. Haugaard et al. and Burashnikov et al. examined the effects of NS8593, along with another SK channel inhibitor (UCL1684), in human and equine atrial myocytes, and verified the ability of both inhibitors to reduce AF inducibility or terminate induced AF [97, 98]. However, conflicting results were found by Fenner et al. in their horse model of tachypacing-induced persistent AF [99]. There, delayed right atrial conduction after NS8593 treatment actually increased AF complexity through increased anisotropy and electrical dissociation [99]. Meanwhile, the left atrium exhibited no change at all in AF complexity, and neither left nor right atrium ultimately resulted in cardioversion [99].

Moving on, treatment with SK2 inhibitor AP30663 in atrially tachypaced live AF pigs resulted in conversion to sinus rhythm, increased right atrial effective refractory periods, and prevented reinduction of AF [100]. Additional studies using whole-cell and inside-out patch clamp recordings of guinea-pig heart tissue confirmed a right-shift of the calcium-activation curve of SK2 channels in the presence of AP30663, with concentration-dependent prolongation of atrial refractoriness and minor effects of QT prolongation [91].

Finally, Saljic et al. tested the use of an antisense oligonucleotide GapmeR in rats and showed that GapmeR downregulates SK3 protein expression in the heart and provides protection against AF [101]. Though targeting the expression level of SK channel seems promising, Darkow et al., showed that that SK3 was upregulated in ventricular tissue in heart failure patients, suggesting that SK channels are not likely to be an atria-selective target as previously expected [25].

In a recent study, Gatta et al. conducted a detailed examination of the effects of SK channel inhibitor AP14145 on goat hearts induced to AF after 30 days of burst-pacing stimulation delivered by pericardial electrodes implanted above the left atria [102]. The authors found that AP14145 produced dose-dependent prolongation of AF cycle length and increased the effective refractory period of atrial impulses [102]. Interestingly, atrial conduction velocity in AF following AP14145 treatment remained unchanged until the final seconds before AF termination, where sudden organization of fibrillatory conduction occurred prior to AF cardioversion [102].

Most animal models of AF discussed up to this point involve AF induction via simple electrode-delivered burst-pacing to atrial tissue. However, other approaches also exist. For example, Yan et al. studied an atrial stretch-induced rabbit AF model [103]. As discussed earlier, atrial enlargement increases the risk of AF by shortening atrial effective refractory periods. Hence the authors of this study placed an inflatable balloon into their rabbit heart left atria to mechanically dilate atria to various sizes. Sustained AF was induced by brief delivery of burst pacing to dilated atria that produced rapid irregular atrial rhythms. For the experimental group, the SK inhibitor ICA was applied to rabbit hearts before atrial stretch and burst pacing. Final analyses showed that ICA pretreatment significantly attenuated stretch-induced atrial effective refractory period and reduced overall AF inducibility and duration when compared with untreated hearts [103]. Another alternative approach was taken by Celotto et al., who focused on the role of autonomic dysfunction in AF pathogenesis [104]. Using human atrial cell and tissue models, the authors induced AF via high dose acetylcholine administration, which shortened atrial APD [104]. SK channel blockade was able to partially revert APD shortening due to acetylcholine, while a combination of SK blockade and the adrenergic agonist isoproterenol were able to completely reverse APD shortening back to pre-AF baseline [104].

How SK channel inhibitors compare to current mainstay antiarrhythmic medications is another important question with significant clinical implications. Two studies by Kirchhoff et al. provides some insights into this issue [105, 106].

In one study, Kirchhoff et al. explored the effect of combining SK channel inhibition via ICA and voltage-gated sodium channel inhibition on AF in atrial-burst pacing induced AF guinea pigs [105]. Ultimately, AF combining ICA with normally subefficacious concentrations of flecainide or ranolazine (sodium channel blockers) reduced AF duration [105].

Next, Kirchhoff et al. examined the effect of the SK inhibitor ICA on AF in an atrial burst-pacing guinea pig model when used alongside amiodarone or dofetilide, two major class III antiarrhythmics. The authors found that combining ICA with either dofetilite or amiodarone reduced AF duration at normally sub-optimal concentrations of all three drugs if used individually [106]. In addition, ICA combined with a standard therapeutic dose of dofetilide prevented QT prolongation that is often seen with dofetilide monotherapy [106]. Overall, both studies suggest that combining SK channel inhibitors with traditional antiarrhythmics may provide a useful synergistic benefit that allows for reducing doses of traditional antiarrhythmics. This in turn may help mitigate against adverse effects of high-dose antiarrhythmics (e.g. long QT, VA).

Note that when all available evidence (pharmacologic, omic, and knockout studies) is considered together, over and underactivity of SK channels both appear to increase likelihood of developing AF. Perhaps discrepancies between different studies are the result of differences in experimental techniques or different species. Alternatively, a two-fold mechanism may drive SK pathology in AF. On one hand, gradually increasing potassium currents due to increased insertion or presence of SK channel in cell membranes may contribute to the observed accelerated actional potential refractory periods and steady progression of APD shortening. On the other hand, action potential prolongation due to diminished potassium currents in the absence of SK channel activity may increase likelihood of generating early afterdepolarizations that may evolve into ectopic foci or reentry circuits. If correct, this proposal implies that seeking to modulate SK channel activity in AF is not a simple matter of complete pharmacologic blockade or complete potentiation. Rather, the objective would be restoring balance towards a homeostatic level of SK channel activity. Further research is required to define the precise pathophysiology of altered SK channel activity in AF to better guide development and protocols for new therapeutic tactics.

5.2 SK channels in ventricular arrhythmias and heart failure

Ventricular arrhythmias (VA) are abnormal heart rhythms that include premature ventricular complexes (PVC), non-sustained ventricular tachycardia (VT), accelerated idioventricular rhythm, and sustained VT or ventricular fibrillation (VF). VT and VF in particular may cause sudden cardiac death (SCD) [107]. Ventricular arrhythmias can be seen in patients with structurally normal hearts, but malignant ventricular arrhythmias usually occur in patients with underlying structural heart disease including HF, ischemic cardiomyopathy, and nonischemic cardiomyopathy [107, 108].

In the normal heart, SK1 channels are expressed at a higher level in atria versus in ventricles, while SK2 and SK3 are expressed at a similar level in both chambers [14, 25]. Normally, apamin-sensitive currents are more prominent in atria than ventricles [12, 14]. However, under certain pathological conditions such as HF and chronic MI, SK currents are upregulated in ventricles, suggesting that SK channels play an important role in ventricular repolarization and VA in pathologic hearts [39, 40].

In healthy ventricles, apamin does not alter APD [14, 22]. Under pathological conditions such as HF, chronic MI, cardiac hypertrophy, and hypokalemia, SK channel blockers prolong APD in ventricles as shown in human and animal models [29, 39, 40, 105, 109]. In acute MI, however, the results are mixed. Some studies show that SK channel blockade prolongs APD in rat acute MI models [110, 111], but a recent study using a porcine model of acute MI showed no significant effect of APD alteration by SK blockade [112]. In the studies that showed prolonged APD by SK blockade in pathologic ventricles, the effects could be either antiarrhythmic or proarrhythmic, probably due to different baseline heart rhythms in the specific animal models used for the studies [18].

Antiarrhythmic effects of SK blockade may occur through attenuation of APD shortening and reduction of repolarization heterogeneity in pathologic hearts [112]. Indeed, Chua et al. were the first to show proarrhythmic effects of SK channels in the ventricles. They showed that HF heterogeneously increased the SK channel's sensitivity to intracellular Ca^{2+} and upregulated SK currents, which led to APD shortening and recurrent spontaneous VF in a rabbit model of tachycardia-induced HF [39]. In this scenario, the rapid heart rate in heart failure caused APD shortening, and excessive APD shortening was arrhythmogenic. APD shortening led to increased intracellular Ca^{2+}, which activated SK currents and further shortened APD, resulting in late phase 3 early afterdepolarization and recurrent spontaneous VF [39]. The antiarrhythmic effects of SK blockers in the ventricles have also been demonstrated in human HF [29], in rabbits with chronic MI [40], in rats with acute MI [110, 111], in rats with cardiac hypertrophy [113, 114], and in hypokalemic guinea pig heart [100].

Chen et al. reported that SK channels are proarrhythmic and play a role in inducing J wave syndrome (JWS). They showed that concurrent activation of SK currents and inhibition of Na^{+} currents shortened APD and induced JWS and SVF in rabbit hearts. SK channel blockade in this rabbit model was antiarrhythmic—JWS was reduced and SVF was abolished, suggesting that SK current activation contributed to the development JWS and SVF in rabbit ventricles [109]. A recent study showed that colocalization of LTCC and SK channels in ventricular myocytes activates SK currents, which then promote phase 2 reentry and T-wave alternans, leading to JWS and VA [115].

Although SK blockers have potential antiarrhythmic benefits in the management of ventricular arrhythmias, blocking SK channels may carry significant proarrhythmic risk in patients with underlying heart disease such as HF, MI, and cardiac hypertrophy, based on animal model studies [116–118]. Blocking SK channels might reduce the repolarization reserve in patients, trigger early after depolarizations (EADs), increase risk of developing torsade's de pointes (TdP) and induce fatal arrhythmia [18, 112, 116]. In addition, SK blockers might be proarrhythmic in hypokalemic hearts. Chan et al. reported that hypokalemia activates SK channels, shortening APD and maintaining the repolarization reserve at late activation sites. In their rabbit model of hypokalemic ventricles, apamin was proarrhythmic by prolonging APD at late activation sites and inducing VF [119].

In addition, Wan et al. showed that SK blockade might even interfere with ventricular automaticity in normal ventricles [120]. Although SK channels do not participate in repolarization in healthy ventricles, SK currents and SK2 protein are prominent in Purkinje cells in normal rabbit ventricles [42]. Wan et al. reported that apamin accelerated ventricular escape rhythms from the Purkinje fibers, enhanced ventricular automaticity and led to VT in normal rabbit ventricles [120].

To summarize, the heterogenous activation of SK channels is proarrhythmic and contributes to the development of ventricular arrythmia in diseased hearts. SK blockers such as apamin have some antiarrhythmic benefits, but also carry significant proarrhythmic risks, thus limiting the practical use of SK blockers for managing ventricular arrythmia. Furthermore, SK channels are widely expressed in the human body including in the nervous system, so blocking SK channels may have undesired off-target effects. Alternatively, drugs that target the signaling pathway of SK channels might have antiarrhythmic effects without directly blocking the channel. Given that increased intracellular Ca^{2+} triggers the upregulation of SK channel in pathologic ventricles, drugs that affect the interactions of SK channel and intracellular Ca^{2+} might have antiarrhythmic or proarrhythmic effects. For example, β-blockers, a known treatment for ventricular arrhythmias, downregulate SK1 and SK3 expression, the SK channel's sensitivity to Ca^{2+}, and the SK current density as shown in a volume-overload rat model [121]. Kamada et al. showed that β- adrenoreceptor stimulation activated SK channels via CaMKII activity in hypertrophied rat hearts, which might contribute to the antiarrhythmic effects of β-blockers [122]. Finally, recent studies showed that sex differences existed with respect to ventricular SK channel activation in response to autonomic stimulation [109, 123], which might have important clinical implications for drug efficacy and safety.

5.3 SK channels and ischemia/reperfusion

During cell ischemia and hypoxia, the altered redox state leads to excess reactive oxygen species (ROS) production which overwhelms the ROS scavenger system, causing mitochondrial Ca^{2+} overload that results in cell apoptosis and necrosis [124]. Given that mitochondria are important in ROS production, ion channels present on the mitochondrial membranes could contribute to the regulation of homeostasis by regulating ROS production during ischemia [124]. In 2013, Stowe et al. discovered that SK channels were located in the guinea pig cardiac IMM, and that SK channel opening had a protective effect during ischemia [37]. In their study, the SK and IK channel opener DCEB resulted in decreased infarct size, reduced superoxide (O_2^-) and mitochondrial Ca^{2+} levels, and more normal NADH and FAD levels. The protective effects were reduced when TBAP, a dismutator of O_2^- was added, suggesting that the benefits of channel openers were related to ROS production [37].

Later on, in a separate study, Stowe et al. also showed that SK channel opening improved contractility and reduced infarct size during ischemia/reperfusion (IR) [125]. In their guinea pig heart model of global IR injury, the SK channel opener DCEB improved contractile function, while SK antagonists worsened contractility and increased infarct size. Furthermore, in cardiac mitochondria after IR, combined SK channel and large conductance Ca^{2+}-activated K^+ (BK) channel agonists improved respiratory control index and Ca^{2+} retention capacity, while the combined antagonists worsened Ca^{2+} retention capacity [125]. Once again, these results show that SK channel plays a role in regulating homeostasis of mitochondria and reducing cell damage during IR.

In 2017, Yang et al., showed that SK3 channels were located in the mitochondria of guinea pig, rat and human ventricular myocytes. They reported that SK channel agonists were protective against IR injury while SK antagonists worsened IR injury. Overexpression of SK3.1 specifically increased Ca^{2+}-activated K^+ uptake in mouse atrial tumor cells and protected the cells from hypoxia/reoxygenation injury. Consistently, silencing SK3.1 channel expression exacerbated cell injury and death [38]. Hence the authors conclude that the protective effect of SK channels during IR suggests their role in reducing oxidative stress resulting from mitochondrial Ca^{2+} overload [38].

In addition, in hypertrophic hearts, mitochondrial SK channels also appear to have protective benefits by decreasing mitochondrial ROS production as shown in a rat model [126]. Kim et al., reported that SK channel enhancers reversed the oxidation of RyRs, improved RyR function and stabilized SR Ca^{2+} release, leading to the protective effects of SK channels in hypertrophic heart [126]. To summarize, mitochondrial SK channel have important protective effects during cardiac cell ischemia, hypoxia and hypertrophy by regulating Ca^{2+} homeostasis and ROS production in the mitochondria.

6. Conclusion

It is increasingly evident that SK channels have important roles in myocardial physiology and pathophysiology. While different studies report different expression levels of various SK channel isoforms in atrial vs. ventricular tissue, all SK channel isoforms present in the heart are necessary for promoting atrial and ventricular repolarization following myocardial impulse generation. The high calcium sensitivity of SK channels, mediated by channel-bound calmodulin and modulated by important regulators such CK2 and PP2A, renders them crucial for feedback control of myocardial contractility and prevention of runaway excitation. In the context of arrythmias, SK channel polymorphisms confer increased risk of atrial fibrillation, and both hyper- and hypoactivity of SK channels along with aberrant SK channel expression likely contribute to automaticity, re-entry circuits, and ectopic pacemaker activity that drives atrial and ventricular tachyarrythmias; these effects are exacerbated in the context of underlying heart disease, such as congestive heart failure. Furthermore, SK channel activity appears to have a protective effect in mitigating oxidative stress during ischemia, with particular significance given to myocardial mitochondrial SK channels.

Pharmacologic studies of SK channel inhibition or activation show promise for treating many animal models of arrythmias and ischemia-reperfusion, although results are still not consistent across different models and different protocols; hence additional research will be required prior to clinical trials of SK channel antagonists in humans. In the future, more research on SK channel pathology using human atrial or ventricular tissue will also be necessary to complement and verify cellular/molecular findings from animal models. Likewise, the specific mechanisms behind altered SK channel expression in many of the diseases discussed in this chapter remain murky and must be elucidated to better characterize specific aspects of pathology at play.

Author details

Katherine Zhong, Shawn Kant, Frank Sellke and Jun Feng*
Division of Cardiothoracic Surgery, Department of Surgery, Rhode Island Hospital, Alpert Medical School of Brown University, Providence, Rhode Island, USA

*Address all correspondence to: jfeng@lifespan.org

References

[1] Grant AO. Cardiac ion channels. Circulation. Arrhythmia and Electrophysiology. 2009;**2**(2):185-194

[2] Amin AS, Tan HL, Wilde AAM. Cardiac ion channels in health and disease. Heart Rhythm. 2010;**7**(1): 117-126

[3] Rahm A-K, Lugenbiel P, Schweizer PA, Katus HA, Thomas D. Role of ion channels in heart failure and channelopathies. Biophysical Reviews. 2018;**10**(4):1097-1106

[4] Wei X, Yohannan S, Richards JR. Physiology, cardiac repolarization dispersion and reserve. In: StatPearls [Internet]. Treasure Island (FL): StatPearls Publishing; 2022. Available from: http://www.ncbi.nlm.nih.gov/books/NBK537194/

[5] Hancock JM, Weatherall KL, Choisy SC, James AF, Hancox JC, Marrion NV. Selective activation of heteromeric SK channels contributes to action potential repolarization in mouse atrial myocytes. Heart Rhythm. 2015; **12**(5):1003-1015

[6] Gu M, Zhu Y, Yin X, Zhang D-M. Small-conductance Ca2+−activated K+ channels: Insights into their roles in cardiovascular disease. Experimental & Molecular Medicine. 2018;**50**(4):1-7

[7] Skibsbye L, Poulet C, Diness JG, Bentzen BH, Yuan L, Kappert U, et al. Small-conductance calcium-activated potassium (SK) channels contribute to action potential repolarization in human atria. Cardiovascular Research. 2014; **103**(1):156-167

[8] Weisbrod D. Small and intermediate calcium activated potassium channels in the heart: Role and strategies in the treatment of cardiovascular diseases. Frontiers in Physiology. 2020;**11**:590534

[9] Adelman JP, Maylie J, Sah P. Small-conductance Ca2+−activated K+ channels: Form and function. Annual Review of Physiology. 2012;**74**:245-269

[10] Köhler M, Hirschberg B, Bond CT, Kinzie JM, Marrion NV, Maylie J, et al. Small-conductance, calcium-activated potassium channels from mammalian brain. Science. 1996;**273**(5282):1709-1714

[11] Stocker M. Ca(2+)-activated K+ channels: Molecular determinants and function of the SK family. Nature Reviews. Neuroscience. 2004;**5**(10): 758-770

[12] Tuteja D, Xu D, Timofeyev V, Lu L, Sharma D, Zhang Z, et al. Differential expression of small-conductance Ca2+−activated K+ channels SK1, SK2, and SK3 in mouse atrial and ventricular myocytes. American Journal of Physiology. Heart and Circulatory Physiology. 2005;**289**(6):H2714-H2723

[13] Wei AD, Gutman GA, Aldrich R, Chandy KG, Grissmer S, Wulff H. International Union of Pharmacology. LII. Nomenclature and molecular relationships of calcium-activated potassium channels. Pharmacological Reviews. 2005;**57**(4):463-472

[14] Xu Y, Tuteja D, Zhang Z, Xu D, Zhang Y, Rodriguez J, et al. Molecular identification and functional roles of a Ca(2+)-activated K+ channel in human and mouse hearts. The Journal of Biological Chemistry. 2003;**278**(49): 49085-49094

[15] Xu Y, Zhang Z, Timofeyev V, Sharma D, Xu D, Tuteja D, et al. The effects of intracellular Ca2+ on cardiac

K+ channel expression and activity: Novel insights from genetically altered mice. The Journal of Physiology. 2005; **562**(Pt 3):745-758

[16] Zhang X-D, Lieu DK, Chiamvimonvat N. Small-conductance Ca2+−activated K+ channels and cardiac arrhythmias. Heart Rhythm. 2015;**12**(8): 1845-1851

[17] Zhang X-D, Thai PN, Lieu DK, Chiamvimonvat N. Cardiac small-conductance calcium-activated potassium channels in health and disease. Pflügers Archiv. 2021;**473**(3): 477-489

[18] Chang P-C, Chen P-S. SK channels and ventricular arrhythmias in heart failure. Trends in Cardiovascular Medicine. 2015;**25**(6):508-514

[19] Feng J, Liu Y, Clements RT, Sodha NR, Khabbaz KR, Senthilnathan V, et al. Calcium-activated potassium channels contribute to human coronary microvascular dysfunction after cardioplegic arrest. Circulation. 2008;**118**(Suppl. 14):S46-S51

[20] Liu Y, Xie A, Singh AK, Ehsan A, Choudhary G, Dudley S, et al. Inactivation of endothelial small/intermediate conductance of calcium-activated potassium channels contributes to coronary arteriolar dysfunction in diabetic patients. Journal of the American Heart Association. 2015; **4**(8):e002062

[21] Liu Y, Cole V, Lawandy I, Ehsan A, Sellke FW, Feng J. Decreased coronary arteriolar response to KCa channel opener after cardioplegic arrest in diabetic patients. Molecular and Cellular Biochemistry. 2018;**445**(1):187-194

[22] Nagy N, Szuts V, Horváth Z, Seprényi G, Farkas AS, Acsai K, et al. Does small-conductance calcium-activated potassium channel contribute to cardiac repolarization? Journal of Molecular and Cellular Cardiology. 2009;**47**(5):656-663

[23] Qi X-Y, Diness JG, Brundel BJJM, Zhou X-B, Naud P, Wu C-T, et al. Role of small-conductance calcium-activated potassium channels in atrial electrophysiology and fibrillation in the dog. Circulation. 2014;**129**(4):430-440

[24] Hsueh C-H, Chang P-C, Hsieh Y-C, Reher T, Chen P-S, Lin S-F. Proarrhythmic effect of blocking the small conductance calcium activated potassium channel in isolated canine left atrium. Heart Rhythm. 2013;**10**(6): 891-898

[25] Darkow E, Nguyen TT, Stolina M, Kari FA, Schmidt C, Wiedmann F, et al. Small conductance Ca2 +−activated K+ (Sk) channel mRNA expression in human atrial and ventricular tissue: Comparison between donor, atrial fibrillation and heart failure tissue. Frontiers in Physiology. 2021;**12**:650964

[26] Ling T-Y, Wang X-L, Chai Q, Lau T-W, Koestler CM, Park SJ, et al. Regulation of the SK3 channel by microRNA-499–potential role in atrial fibrillation. Heart Rhythm. 2013;**10**(7):1001-1009

[27] Yu T, Deng C, Wu R, Guo H, Zheng S, Yu X, et al. Decreased expression of small-conductance Ca2+−activated K+ channels SK1 and SK2 in human chronic atrial fibrillation. Life Sciences. 2012;**90**(5–6): 219-227

[28] Hamilton S, Polina I, Terentyeva R, Bronk P, Kim TY, Roder K, et al. PKA phosphorylation underlies functional recruitment of sarcolemmal SK2 channels in ventricular myocytes from hypertrophic hearts. The Journal of Physiology. 2020;**598**(14):2847-2873

[29] Chang P-C, Turker I, Lopshire JC, Masroor S, Nguyen B-L, Tao W, et al. Heterogeneous upregulation of apamin-sensitive potassium currents in failing human ventricles. Journal of the American Heart Association. 2013;**2**(1): e004713

[30] Diness JG, Sørensen US, Nissen JD, Al-Shahib B, Jespersen T, Grunnet M, et al. Inhibition of small-conductance Ca2+-activated k + channels terminates and protects against atrial fibrillation. Circulation. Arrhythmia and Electrophysiology. 2010;**3**(4):380-390

[31] Mahida S, Mills RW, Tucker NR, Simonson B, Macri V, Lemoine MD, et al. Overexpression of KCNN3 results in sudden cardiac death. Cardiovascular Research. 2014;**101**(2):326-334

[32] Fan X, Yu Y, Lan H, Ou X, Yang L, Li T, et al. Ca2+/calmodulin-dependent protein kinase ii (Camkii) increases small-conductance ca2+-activated k+ current in patients with chronic atrial fibrillation. Medical Science Monitor. 2018;**24**:3011-3023

[33] Liu Y, Kabakov AY, Xie A, Shi G, Singh AK, Sodha NR, et al. Metabolic regulation of endothelial SK channels and human coronary microvascular function. International Journal of Cardiology. 2020;**312**:1-9

[34] Ozgen N, Dun W, Sosunov E, Anyukhovsky E, Hirose M, Duffy H, et al. Early electrical remodeling in rabbit pulmonary vein results from trafficking of intracellular SK2 channels to membrane sites. Cardiovascular Research. 2007;**75**(4):758-769

[35] Chen W-T, Chen Y-C, Lu Y-Y, Kao Y-H, Huang J-H, Lin Y-K, et al. Apamin modulates electrophysiological characteristics of the pulmonary vein and the sinoatrial node. European Journal of Clinical Investigation. 2013; **43**(9):957-963

[36] Zhang Q, Timofeyev V, Lu L, Li N, Singapuri A, Long MK, et al. Functional roles of a Ca2+-activated K+ channel in atrioventricular nodes. Circulation Research. 2008;**102**(4):465-471

[37] Stowe DF, Gadicherla AK, Zhou Y, Aldakkak M, Cheng Q, Kwok W-M, et al. Protection against cardiac injury by small Ca2+−sensitive K+ channels identified in Guinea pig cardiac inner mitochondrial membrane. Biochimica et Biophysica Acta (BBA)—Biomembranes. 2013;**1828**(2):427-442

[38] Yang M, Camara AKS, Aldakkak M, Kwok W-M, Stowe DF. Identity and function of a cardiac mitochondrial small conductance Ca2+ −activated K+ channel splice variant. Biochimica et Biophysica Acta (BBA)—Bioenergetics. 2017;**1858**(6):442-458

[39] Chua S-K, Chang P-C, Maruyama M, Turker I, Shinohara T, Shen MJ, et al. Small-conductance calcium-activated potassium channel and recurrent ventricular fibrillation in failing rabbit ventricles. Circulation Research. 2011; **108**(8):971-979

[40] Lee YS, Chang P-C, Hsueh C-H, Maruyama M, Park HW, Rhee K-S, et al. Apamin-sensitive calcium-activated potassium currents in rabbit ventricles with chronic myocardial infarction: SK currents in chronic mi. Journal of Cardiovascular Electrophysiology. 2013; **24**(10):1144-1153

[41] Torrente AG, Zhang R, Wang H, Zaini A, Kim B, Yue X, et al. Contribution of small conductance K+ channels to sinoatrial node pacemaker activity: Insights from atrial-specific Na+/Ca2+ exchange knockout mice: Role of small conductance K + channels in

sinus node activity. The Journal of Physiology. 2017;**595**(12):3847-3865

[42] Reher TA, Wang Z, Hsueh C-H, Chang P-C, Pan Z, Kumar M, et al. Small-conductance calcium-activated potassium current in normal rabbit cardiac purkinje cells. Journal of the American Heart Association. 2017;**6**(6):e005471

[43] Bond CT. Small conductance Ca2+−activated K+ channel knock-out mice reveal the identity of calcium-dependent afterhyperpolarization currents. Journal of Neuroscience. 2004;**24**(23):5301-5306

[44] Zhang X-D, Timofeyev V, Li N, Myers RE, Zhang D-M, Singapuri A, et al. Critical roles of a small conductance Ca2+−activated K+ channel (Sk3) in the repolarization process of atrial myocytes. Cardiovascular Research. 2014;**101**(2):317-325

[45] Action potential—en.ecgpedia.org. n.d. Available from: https://en.ecgpedia.org/wiki/Action_potential [Accessed: February 18, 2022]

[46] Marrion NV, Tavalin SJ. Selective activation of Ca2+-activated K+ channels by co-localized Ca2+ channels in hippocampal neurons. Nature. 1998; **395**(6705):900-905

[47] Lu L, Zhang Q, Timofeyev V, Zhang Z, Young JN, Shin H-S, et al. Molecular coupling of a Ca2+-activated K+ channel to L-type Ca2+ channels via α-actinin2. Circulation Research. 2007; **100**(1):112-120

[48] Mu Y, Zhao W, Duan P, Chen Y, Zhao W, Wang Q, et al. Ryr2 modulates a Ca2+−activated K+ current in mouse cardiac myocytes. Attali B, editor. PLoS One. 2014;**9**(4):e94905

[49] Xia X-M, Fakler B, Rivard A, Wayman G, Johnson-Pais T, Keen JE, et al. Mechanism of calcium gating in small-conductance calcium-activated potassium channels. Nature. 1998; **395**(6701):503-507

[50] Schumacher MA, Rivard AF, Bächinger HP, Adelman JP. Structure of the gating domain of a Ca2+-activated K+ channel complexed with Ca2+/calmodulin. Nature. 2001;**410**(6832): 1120-1124

[51] Maylie J, Bond CT, Herson PS, Lee W-S, Adelman JP. Small conductance Ca2+-activated K+ channels and calmodulin: SK channels and calmodulin. The Journal of Physiology. 2004;**554**(2): 255-261

[52] Lu L, Timofeyev V, Li N, Rafizadeh S, Singapuri A, Harris TR, et al. Actinin2 cytoskeletal protein is required for the functional membrane localization of a Ca2+-activated K+ channel (SK2 channel). Proceedings of the National Academy of Sciences. 2009; **106**(43):18402-18407

[53] Lu L, Sirish P, Zhang Z, Woltz RL, Li N, Timofeyev V, et al. Regulation of gene transcription by voltage-gated l-type calcium channel, cav1. 3. Journal of Biological Chemistry. 2015;**290**(8): 4663-4676

[54] Rafizadeh S, Zhang Z, Woltz RL, Kim HJ, Myers RE, Lu L, et al. Functional interaction with filamin A and intracellular Ca2+ enhance the surface membrane expression of a small-conductance Ca2+-activated K+ (Sk2) channel. Proceedings of the National Academy of Sciences of the United States of America. 2014;**111**(27): 9989-9994

[55] Yang D, Wang T, Ni Y, Song B, Ning F, Hu P, et al. Apamin-sensitive K+ current upregulation in volume-overload heart failure is associated with the

decreased interaction of CK2 with SK2. Journal of Membrane Biology. 2015; **248**(6):1181-1189

[56] Mizukami K, Yokoshiki H, Mitsuyama H, Watanabe M, Tenma T, Takada S, et al. Small-conductance Ca^{2+}-activated K + current is upregulated via the phosphorylation of CaMKII in cardiac hypertrophy from spontaneously hypertensive rats. American Journal of Physiology-Heart and Circulatory Physiology. 2015;**309**(6):H1066-H1074

[57] Pellman J, Sheikh F. Atrial fibrillation: Mechanisms, therapeutics, and future directions. Comprehensive Physiology. 2015;**5**(2):649-665

[58] Brandes A, Smit MD, Nguyen BO, Rienstra M, Van Gelder IC. Risk factor management in atrial fibrillation. Arrhythmia & Electrophysiology Review. 2018;**7**(2):118-127

[59] Andrade J, Khairy P, Dobrev D, Nattel S. The clinical profile and pathophysiology of atrial fibrillation: Relationships among clinical features, epidemiology, and mechanisms. Circulation Research. 2014;**114**(9): 1453-1468

[60] January CT, Wann LS, Alpert JS, Calkins H, Cigarroa JE, Cleveland JC, et al. 2014 aha/acc/hrs guideline for the management of patients with atrial fibrillation. Journal of the American College of Cardiology. 2014;**64**(21):e1-e76

[61] Ellinor PT, Lunetta KL, Glazer NL, Pfeufer A, Alonso A, Chung MK, et al. Common variants in KCNN3 are associated with lone atrial fibrillation. Nature Genetics. 2010;**42**(3):240-244

[62] Nattel S, Dobrev D. The multidimensional role of calcium in atrial fibrillation pathophysiology: Mechanistic insights and therapeutic opportunities. European Heart Journal. 2012;**33**(15):1870-1877

[63] Pandit SV, Berenfeld O, Anumonwo JMB, Zaritski RM, Kneller J, Nattel S, et al. Ionic determinants of functional reentry in a 2-d model of human atrial cells during simulated chronic atrial fibrillation. Biophysical Journal. 2005;**88**(6):3806-3821

[64] Voigt N, Maguy A, Yeh Y-H, Qi X, Ravens U, Dobrev D, et al. Changes in IK,ACh single-channel activity with atrial tachycardia remodelling in canine atrial cardiomyocytes. Cardiovascular Research. 2007;**77**(1):35-43

[65] Igarashi T, Finet JE, Takeuchi A, Fujino Y, Strom M, Greener ID, et al. Connexin gene transfer preserves conduction velocity and prevents atrial fibrillation. Circulation. 2012;**125**(2): 216-225

[66] Dun W, Boyden PA. Aged atria: Electrical remodeling conducive to atrial fibrillation. Journal of Interventional Cardiac Electrophysiology. 2009;**25**(1): 9-18

[67] Nishida K, Qi XY, Wakili R, Comtois P, Chartier D, Harada M, et al. Mechanisms of atrial tachyarrhythmias associated with coronary artery occlusion in a chronic canine model. Circulation. 2011;**123**(2):137-146

[68] Deroubaix E, Folliguet T, Rücker-Martin C, Dinanian S, Boixel C, Validire P, et al. Moderate and chronic hemodynamic overload of sheep atria induces reversible cellular electrophysiologic abnormalities and atrial vulnerability. Journal of the American College of Cardiology. 2004; **44**(9):1918-1926

[69] Kim S, Iwao H. Molecular and cellular mechanisms of angiotensin

II-mediated cardiovascular and renal diseases. Pharmacological Reviews. 2000;**52**(1):11-34

[70] Tsai C-T, Chiang F-T, Tseng C-D, Hwang J-J, Kuo K-T, Wu C-K, et al. Increased expression of mineralocorticoid receptor in human atrial fibrillation and a cellular model of atrial fibrillation. Journal of the American College of Cardiology. 2010; **55**(8):758-770

[71] Burstein B, Comtois P, Michael G, Nishida K, Villeneuve L, Yeh Y-H, et al. Changes in connexin expression and the atrial fibrillation substrate in congestive heart failure. Circulation Research. 2009; **105**(12):1213-1222

[72] Zlochiver S, Muñoz V, Vikstrom KL, Taffet SM, Berenfeld O, Jalife J. Electrotonic myofibroblast-to-myocyte coupling increases propensity to reentrant arrhythmias in two-dimensional cardiac monolayers. Biophysical Journal. 2008;**95**(9): 4469-4480

[73] Zou R, Kneller J, Leon LJ, Nattel S. Substrate size as a determinant of fibrillatory activity maintenance in a mathematical model of canine atrium. American Journal of Physiology-Heart and Circulatory Physiology. 2005; **289**(3):H1002-H1012

[74] Sharifov OF, Fedorov VV, Beloshapko GG, Glukhov AV, Yushmanova AV, Rosenshtraukh LV. Roles of adrenergic and cholinergic stimulation in spontaneous atrial fibrillation in dogs. Journal of the American College of Cardiology. 2004; **43**(3):483-490

[75] Chang C-M, Wu T-J, Zhou S, Doshi RN, Lee M-H, Ohara T, et al. Nerve sprouting and sympathetic hyperinnervation in a canine model of atrial fibrillation produced by prolonged right atrial pacing. Circulation. 2001; **103**(1):22-25

[76] Shen MJ, Chang H-C, Park H-W, George Akingba A, Chang P-C, Zhang Z, et al. Low-level vagus nerve stimulation upregulates small conductance calcium-activated potassium channels in the stellate ganglion. Heart Rhythm. 2013; **10**(6):910-915

[77] Yu L, Scherlag BJ, Li S, Sheng X, Lu Z, Nakagawa H, et al. Low-level vagosympathetic nerve stimulation inhibits atrial fibrillation inducibility: Direct evidence by neural recordings from intrinsic cardiac ganglia. Journal of Cardiovascular Electrophysiology. 2011; **22**(4):455-463

[78] Lubitz SA, Ozcan C, Magnani JW, Kääb S, Benjamin EJ, Ellinor PT. Genetics of atrial fibrillation: Implications for future research directions and personalized medicine. Circulation. Arrhythmia and Electrophysiology. 2010;**3**(3):291-299

[79] Xu C, Zhang R, Xia Y, Xiong L, Yang W, Wang P. Annotation of susceptibility SNPs associated with atrial fibrillation. Aging. 2020;**12**(17): 16981-16998

[80] Feghaly J, Zakka P, London B, MacRae CA, Refaat MM. Genetics of atrial fibrillation. Journal of the American Heart Association. 2018;7(20): e009884. DOI: 10.1161/ JAHA.118.009884

[81] Chen S, Zhang L, Bryant R, Vincent G, Flippin M, Lee J, et al. kcnq1 mutations in patients with a family history of lethal cardiac arrhythmias and sudden death: Genotype-phenotype correlations in patients with kcnq1 mutations. Clinical Genetics. 2003; **63**(4):273-282

[82] Sinner MF, Pfeufer A, Akyol M, Beckmann B-M, Hinterseer M, Wacker A, et al. The non-synonymous coding IKr-channel variant KCNH2-K897T is associated with atrial fibrillation: Results from a systematic candidate gene-based analysis of KCNH2 (Herg). European Heart Journal. 2008; **29**(7):907-914

[83] Olesen MS, Refsgaard L, Holst AG, Larsen AP, Grubb S, Haunsø S, et al. A novel KCND3 gain-of-function mutation associated with early-onset of persistent lone atrial fibrillation. Cardiovascular Research. 2013;**98**(3):488-495

[84] Xia M, Jin Q, Bendahhou S, He Y, Larroque M-M, Chen Y, et al. A Kir2.1 gain-of-function mutation underlies familial atrial fibrillation. Biochemical and Biophysical Research Communications. 2005;**332**(4): 1012-1019

[85] Olson TM, Alekseev AE, Liu XK, Park S, Zingman LV, Bienengraeber M, et al. Kv1.5 channelopathy due to KCNA5 loss-of-function mutation causes human atrial fibrillation. Human Molecular Genetics. 2006;**15**(14):2185-2191

[86] Chang S-H, Chang S-N, Hwang J-J, Chiang F-T, Tseng C-D, Lee J-K, et al. Significant association of rs13376333 in kcnn3 on chromosome 1q21 with atrial fibrillation in a Taiwanese population. Circulation Journal. 2012;**76**(1):184-188

[87] Luo Z, Yan C, Zhang W, Shen X, Zheng W, Chen F, et al. Association between SNP rs13376333 and rs1131820 in the KCNN3 gene and atrial fibrillation in the Chinese Han population. Clinical Chemistry and Laboratory Medicine. 2014;**52**(12):1867-1873

[88] Li C, Wang F, Yang Y, Fu F, Xu C, Shi L, et al. Significant association of SNP rs2106261 in the ZFHX3 gene with atrial fibrillation in a Chinese Han GeneID population. Human Genetics. 2011;**129**(3):239-246

[89] Olesen MS, Jabbari J, Holst AG, Nielsen JB, Steinbrüchel DA, Jespersen T, et al. Screening of KCNN3 in patients with early-onset lone atrial fibrillation. EP Europace. 2011;**13**(7):963-967

[90] Yu C-C, Chia-Ti T, Chen P-L, Wu C-K, Chiu F-C, Chiang F-T, et al. KCNN2 polymorphisms and cardiac tachyarrhythmias. Medicine. 2016; **95**(29):e4312

[91] Bentzen BH, Bomholtz SH, Simó-Vicens R, Folkersen L, Abildgaard L, Speerschneider T, et al. Mechanisms of action of the kca2-negative modulator ap30663, a novel compound in development for treatment of atrial fibrillation in man. Frontiers in Pharmacology. 2020;**11**:610

[92] Tsai W-C, Chan Y-H, Hsueh C-H, Everett TH, Chang P-C, Choi E-K, et al. Small conductance calcium-activated potassium current and the mechanism of atrial arrhythmia in mice with dysfunctional melanocyte-like cells. Heart Rhythm. 2016;**13**(7):1527-1535

[93] Rahm A-K, Gramlich D, Wieder T, Müller ME, Schoeffel A, El Tahry FA, et al. Trigger-specific remodeling of kca2 potassium channels in models of atrial fibrillation. Pharmacogenomics and Personalized Medicine. 2021;**14**:579-590

[94] Li N, Timofeyev V, Tuteja D, Xu D, Lu L, Zhang Q, et al. Ablation of a Ca 2+ −activated K + channel (SK2 channel) results in action potential prolongation in atrial myocytes and atrial fibrillation: Ablation of SK2 channel results in action potential prolongation in atrial myocytes. The Journal of Physiology. 2009;**587**(5): 1087-1100

[95] Rahm A-K, Wieder T, Gramlich D, Müller ME, Wunsch MN, El Tahry FA, et al. HDAC2-dependent remodeling of KCa2.2 (Kcnn2) and KCa2.3 (Kcnn3) K+ channels in atrial fibrillation with concomitant heart failure. Life Sciences. 2021;**266**:118892

[96] Bauer A, Koch M, Kraft P, Becker R, Kelemen K, Voss F, et al. The new selective IKs? Blocking agent HMR 1556 restores sinus rhythm and prevents heart failure in pigs with persistent atrial fibrillation. Basic Research in Cardiology. 2005;**100**(3):270-278

[97] Haugaard MM, Hesselkilde EZ, Pehrson S, Carstensen H, Flethøj M, Præstegaard KF, et al. Pharmacologic inhibition of small-conductance calcium-activated potassium (Sk) channels by NS8593 reveals atrial antiarrhythmic potential in horses. Heart Rhythm. 2015; **12**(4):825-835

[98] Burashnikov A, Barajas-Martinez H, Hu D, Robinson VM, Grunnet M, Antzelevitch C. The small conductance calcium-activated potassium channel inhibitors ns8593 and ucl1684 prevent the development of atrial fibrillation through atrial-selective inhibition of sodium channel activity. Journal of Cardiovascular Pharmacology. 2020; **76**(2):164-172

[99] Fenner MF, Gatta G, Sattler S, Kuiper M, Hesselkilde EM, Adler DMT, et al. Inhibition of small-conductance calcium-activated potassium current (Ik, ca) leads to differential atrial electrophysiological effects in a horse model of persistent atrial fibrillation. Frontiers in Physiology. 2021;**12**:614483

[100] Diness JG, Abildgaard L, Bomholtz SH, Skarsfeldt MA, Edvardsson N, Sørensen US, et al. Inhibition of kca2 channels decreased the risk of ventricular arrhythmia in the Guinea pig heart during induced hypokalemia. Frontiers in Pharmacology. 2020;**11**:749

[101] Saljic A, Soattin L, Trachsel DS, Boddum K, Jespersen T. In vivo knockdown of SK3 channels using antisense oligonucleotides protects against atrial fibrillation in rats. Journal of Molecular and Cellular Cardiology. 2020;**147**:18-26

[102] Gatta G, Sobota V, Citerni C, Diness JG, Sørensen US, Jespersen T, et al. Effective termination of atrial fibrillation by SK channel inhibition is associated with a sudden organization of fibrillatory conduction. EP Europace. 2021;**23**(11):1847-1859

[103] Yan Y, Skarsfeldt MA, Diness JG, Bentzen BH. Small conductance calcium activated K+ channel inhibitor decreases stretch induced vulnerability to atrial fibrillation. IJC Heart & Vasculature. 2021;**37**:100898

[104] Celotto C, Sanchez C, Mountris KA, Laguna P, Pueyo E. Sk channel block and adrenergic stimulation counteract acetylcholine-induced arrhythmogenic effects in human atria. Annual International Conference of the IEEE Engineering in Medicine and Biology Society. 2020;**2020**:2303-2306

[105] Kirchhoff JE, Goldin Diness J, Sheykhzade M, Grunnet M, Jespersen T. Synergistic antiarrhythmic effect of combining inhibition of Ca2+-activated K+ (Sk) channels and voltage-gated Na+ channels in an isolated heart model of atrial fibrillation. Heart Rhythm. 2015; **12**(2):409-418

[106] Kirchhoff JE, Diness JG, Abildgaard L, Sheykhzade M, Grunnet M, Jespersen T. Antiarrhythmic effect of the Ca2+-activated K+ (Sk) channel inhibitor ICA combined with

either amiodarone or dofetilide in an isolated heart model of atrial fibrillation. Pflügers Archiv—European Journal of Physiology. 2016;**468**(11–12):1853-1863

[107] Bradfield JS, Boyle NG, Shivkumar K. Ventricular arrhythmias. In: Fuster V, Harrington RA, Narula J, Eapen ZJ, editors. Hurst's The Heart [Internet]. 14th ed. New York, NY: McGraw-Hill Education; 2017. Available from: accessmedicine.mhmedical.com/content.aspx?aid=1161718616

[108] Hynes BJ, Luck JC, Wolbrette DL, Boehmer J, Naccarelli GV. Arrhythmias in patients with heart failure. Current Treatment Options in Cardiovascular Medicine. 2002;**4**(6):467-485

[109] Chen M, Xu D-Z, Wu AZ, Guo S, Wan J, Yin D, et al. Concomitant SK current activation and sodium current inhibition cause J wave syndrome. JCI Insight. 2018;**3**(22):e122329

[110] Gui L, Bao Z, Jia Y, Qin X, (Jack) Cheng Z, Zhu J, et al. Ventricular tachyarrhythmias in rats with acute myocardial infarction involves activation of small-conductance Ca2+-activated K + channels. American Journal of Physiology-Heart and Circulatory Physiology. 2013;**304**(1):H118-H130

[111] Hundahl LA, Sattler SM, Skibsbye L, Diness JG, Tfelt-Hansen J, Jespersen T. Pharmacological blockade of small conductance Ca2+-activated K+ channels by ICA reduces arrhythmic load in rats with acute myocardial infarction. Pflügers Archiv—European Journal of Physiology. 2017;**469**(5–6):739-750

[112] Lubberding AF, Sattler SM, Grunnet M, Sørensen US, Tfelt-Hansen J, Jespersen T. Arrhythmia development during inhibition of small-conductance calcium-activated potassium channels in acute myocardial infarction in a porcine model. EP Europace. 2019;**21**(10):1584-1593

[113] Tenma T, Mitsuyama H, Watanabe M, Kakutani N, Otsuka Y, Mizukami K, et al. Small-conductance Ca2+−activated K+ channel activation deteriorates hypoxic ventricular arrhythmias via CaMKII in cardiac hypertrophy. American Journal of Physiology. Heart and Circulatory Physiology. 2018;**315**(2):H262-H272

[114] Takahashi M, Yokoshiki H, Mitsuyama H, Watanabe M, Temma T, Kamada R, et al. SK channel blockade prevents hypoxia-induced ventricular arrhythmias through inhibition of Ca 2+/voltage uncoupling in hypertrophied hearts. American Journal of Physiology-Heart and Circulatory Physiology. 2021;**320**(4):H1456-H1469

[115] Landaw J, Zhang Z, Song Z, Liu MB, Olcese R, Chen P-S, et al. Small-conductance Ca2+−activated K+ channels promote J-wave syndrome and phase 2 reentry. Heart Rhythm. 2020;**17**(9):1582-1590

[116] Chang P-C, Hsieh Y-C, Hsueh C-H, Weiss JN, Lin S-F, Chen P-S. Apamin induces early afterdepolarizations and torsades de pointes ventricular arrhythmia from failing rabbit ventricles exhibiting secondary rises in intracellular calcium. Heart Rhythm. 2013;**10**(10):1516-1524

[117] Bonilla IM, Long VP, Vargas-Pinto P, Wright P, Belevych A, Lou Q, et al. Calcium-activated potassium current modulates ventricular repolarization in chronic heart failure. Rodriguez B, editor. PLoS ONE. 2014;**9**(10):e108824

[118] Kim TY, Terentyeva R, Roder KH, Li W, Liu M, Greener I, et al. SK channel enhancers attenuate Ca^{2+}-dependent

arrhythmia in hypertrophic hearts by regulating mito-ROS-dependent oxidation and activity of RyR. Cardiovascular Research. 2017;**113**(3): 343–353. DOI: 10.1093/cvr/cvx005

[119] Chan Y-H, Tsai W-C, Ko J-S, Yin D, Chang P-C, Rubart M, et al. Small-conductance calcium-activated potassium current is activated during hypokalemia and masks short-term cardiac memory induced by ventricular pacing. Circulation. 2015;**132**(15): 1377-1386

[120] Wan J, Chen M, Wang Z, Everett TH, Rubart-von der Lohe M, Shen C, et al. Small-conductance calcium-activated potassium current modulates the ventricular escape rhythm in normal rabbit hearts. Heart Rhythm. 2019;**16**(4):615-623

[121] Ni Y, Wang T, Zhuo X, Song B, Zhang J, Wei F, et al. Bisoprolol reversed small conductance calcium-activated potassium channel (Sk) remodeling in a volume-overload rat model. Molecular and Cellular Biochemistry. 2013;**384** (1–2):95-103

[122] Kamada R, Yokoshiki H, Mitsuyama H, Watanabe M, Mizukami K, Tenma T, et al. Arrhythmogenic β-adrenergic signaling in cardiac hypertrophy: The role of small-conductance calcium-activated potassium channels via activation of CaMKII. European Journal of Pharmacology. 2019;**844**:110-117

[123] Fei Y-D, Chen M, Guo S, Ueoka A, Chen Z, Rubart-von der Lohe M, et al. Simultaneous activation of the small conductance calcium-activated potassium current by acetylcholine and inhibition of sodium current by ajmaline cause J-wave syndrome in Langendorff-perfused rabbit ventricles. Heart Rhythm. 2021;**18**(1):98-108

[124] Stowe DF, Camara AKS. Mitochondrial reactive oxygen species production in excitable cells: Modulators of mitochondrial and cell function. Antioxidants & Redox Signaling. 2009; **11**(6):1373-1414

[125] Stowe DF, Yang M, Heisner JS, Camara AKS. Endogenous and agonist-induced opening of mitochondrial big versus small Ca2+-sensitive K+ channels on cardiac cell and mitochondrial protection. Journal of Cardiovascular Pharmacology. 2017;**70**(5):314-328

[126] Kim TY, Terentyeva R, Roder KHF, Li W, Liu M, Greener I, et al. SK channel enhancers attenuate Ca2+–dependent arrhythmia in hypertrophic hearts by regulating Mito-ROS-dependent oxidation and activity of RyR. Cardiovascular Research. 2017;**113**(3): 343-353

Chapter 6

Vacuolar ATPase (V-ATPase) Proton Pump and Its Significance in Human Health

Anuj Tripathi and Smita Misra

Abstract

Vacuolar H + -ATPases (V-ATPase), is an ATP-dependent proton transporter that transports protons across intracellular and cellular plasma membranes. V-ATPase is a multi-protein complex, which functions as an ATP-driven proton pump and is involved in maintaining pH homeostasis. The V-ATPase is a housekeeping proton pump and is highly conserved during evolution. The proton-pumping activity of V-ATPases allows acidification of intracellular compartments and influences a diverse range of cellular and biological processes. Thus, V-ATPase aberrant overexpression, mis-localization, and mutations in the genes for subunits are associated with several human diseases. This chapter focuses on a detailed view of V-type ATPase, and how V-ATPase contributes to human health and disease.

Keywords: pH, homeostasis, V-ATPases, proton pump, human health

1. Introduction

The maintenance of pH homeostasis is vital for the survival of all cells and organisms. Changes in intracellular pH affect the acid-base balance of the cells, and dictates the protonation state of different acid-base groups present on the macromolecules. This greatly influences their biochemical properties and function. Deregulation of the pH homeostasis affects enzymatic functions affecting the cell cycle and other biochemical processes and can be deleterious for cellular health and survival. In addition to the cytosol, each organelle has its specific pH requirement to function normally. The pH within the cytosol and the organelles can vary up to 3 units ranging from nearly neutral to highly acidic. To maintain the pH all eukaryotic cells, have a large regulatory network of secretory pathways within the cell cytosol and organelles including the nucleus, and outside the plasma membrane [1, 2]. These secretory pathways, from plasma membrane to organelles and nucleus are well connected by continuous exchange of nutrients, signaling molecules, membrane proteins, and lipids. The maintenance and assembly of these complexes and pathways is highly energy consuming for cells. Cellular energy requirements for these processes are partly fulfilled by local cytoplasmic metabolic energy, but a larger extent of the energy required for development and homeostasis maintenance of the cytosol and organelle lumen is provided by ion pumps [3, 4].

Proton pumping ATPases are a class of these membrane transporters that act as master players in the transport of protons across membranes from Archaea to humans. pH homeostasis is achieved *via* proton influx and efflux by the proton pumping ATPases. These ATPases either actively transport proton deriving energy from ATP hydrolysis, or they use the proton gradient for ATP synthesis to perform multiple cellular functions. The ATPases are broadly classified into four classes: the F- and A-type ATP synthases, the V-type transporters, the P-type transporters, and the ATP binding cassette (ABC) multidrug efflux pumps. The A-type ATPase, which are present in archaea and bacteria, help survive the extreme conditions and act by synthesizing ATP coupling with H^+ or Na^+. F-type ATPase are well conserved among species and are the primary source of cellular energy production in the living organisms. They act as ATP synthases. The P-type ATPases, also known as E1–E2 ATPases, are a large group of evolutionarily related ion and lipid pumps that are found in bacteria, archaea, and eukaryotes. The P-ATPase are essential for cell survival, and they maintain the gradient of many crucial ions including Na^+, Ca^+, K^+, and H^+ using ATP hydrolysis. There are different type of P-type ATPases. ABC-transporters utilize the energy of ATP binding and hydrolysis to transport various substrates across cellular membranes. ABC-transporters are importers of nutrients and other molecules, or as exporters of toxins and drugs, among others. V-type ATPase is a highly conserved evolutionarily ancient enzyme with remarkably diverse functions in eukaryotic organisms. They couple ATP-hydrolysis with H^+ pumping.

2. Vacuolar-type ATPase (V-ATPase)

The Vacuolar proton-translocating ATPase (V-ATPase) is a highly conserved and highly efficient ATP driven proton pump and a member of the rotary ATPase protein family [5, 6]. V-ATPase are ubiquitous multi-subunit complexes composed of two large domains: the soluble V_1 domain, which hydrolyzes ATP, and the membrane-embedded V_0 domain, which transports protons [5, 7–9]. V-ATPase were first discovered in vacuoles of yeast and plants. V-ATPase perform active proton transport across membranes by coupling it with ATP hydrolysis. V-ATPase are also identified in the lysosomes, clathrin-coated vesicles, secretory vesicles, endosomes, Golgi-derived vesicles and other subcellular locations. They are also present on the plasma membrane. V-ATPase acidifies lysosomes/vacuoles, Golgi, and the endosomal compartments of all eukaryotes. The plasma membrane V-ATPase present on certain specialized mammalian cells aid in proton export from the cell [10]. In intracellular compartments, V-ATPase is critical for multiple cellular processes, this includes protein processing and secretion, endocytosis and vesicle trafficking, zymogen activation, and autophagy [5, 10]. V-ATPase was initially identified and characterized for its role in the acidification of intracellular vesicles and organelles, which is necessary for many essential cell biological events to occur [9–11]. In addition to its housekeeping cellular function, many specialized cell types in various organ systems such as the kidney, bone, male reproductive tract, inner ear, olfactory mucosa, and others, use plasma membrane V-ATPases to perform specific activities that depend on extracellular acidification [12–16].

Finally, and importantly, it is increasingly apparent that V-ATPases are central players in other normal and pathophysiological processes that directly contribute to human health in many different and sometimes unexpected ways. This chapter will cover the basic knowledge of V-ATPase, its physiological contribution and recently emerging unconventional roles of the V-ATPase in human health.

2.1 Structure and role of V-ATPases

V-ATPase are multi-subunit protein complex with two domains the V_1-domain and V_0-domain. The peripheral domain V_1, is cytosolic and responsible for ATP hydrolysis, and an integral domain V_0, is embedded in the membrane and is involved in proton translocation across the membrane [17]. The mammalian V-ATPase is composed of 13 subunits in total. Of these 13 subunits, V_1 domain has 8 peripheral proteins and the V_0 domain has 5 membrane intrinsic proteins (**Figure 1**) [17]. The V_1 domain performs ATP binding, hydrolysis and drives the active proton translocation from the V_0 domain. Alternate arrangement of V_1A and V_1B subunits forms the hexameric core of the V_1 domain. The V_0 core ring domain is made up of subunits V_0c, V_0c' and V_0c''. The V_0 core ring domain is located next to the V_0a and V_0e subunits. The V_0 and V_1 domains are connected by a central stalk. The central stalk is composed of the V_1D, V_1F and V_0d and is supported by the peripheral stalk domain. The peripheral stalk is made from the subunits V_1C, V_1E, V_1G V_1H and the N-terminal of the V_0a. V_0a is a key subunit of the V_0 domain. It has a bi-lobed N-terminal which interacts with the V_1H and V_1C near the membrane interface and V_1A on the outer surface [18–20]. Arginine at position 735 and two hemi channels of the V_0a subunits are crucial for its proton pumping function.

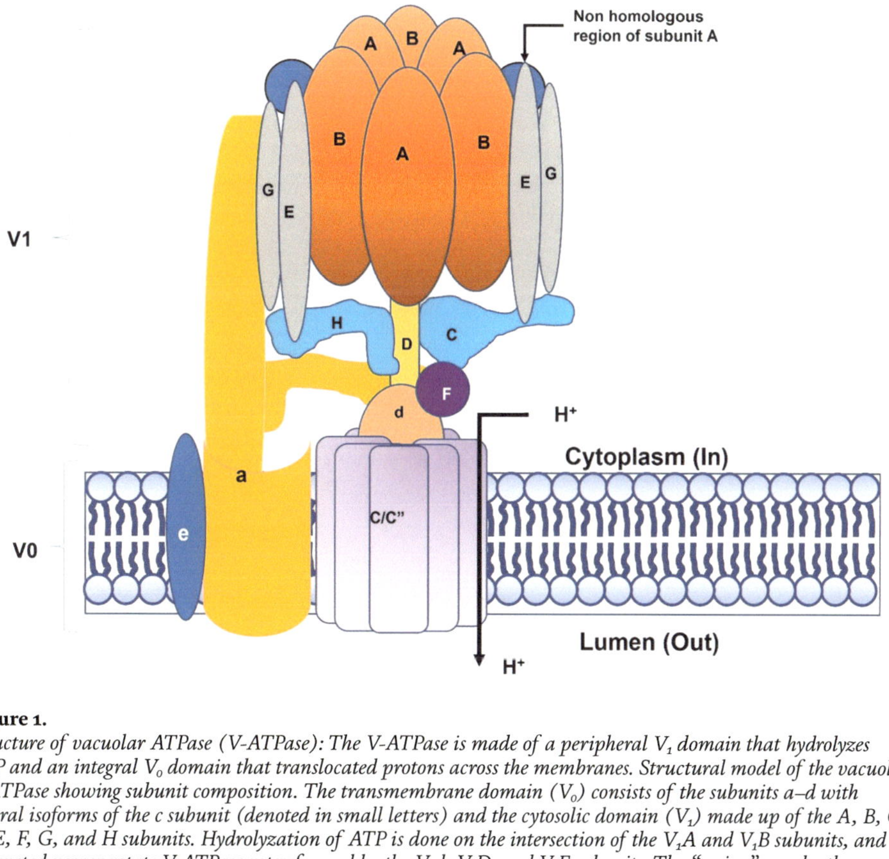

Figure 1.
Structure of vacuolar ATPase (V-ATPase): The V-ATPase is made of a peripheral V_1 domain that hydrolyzes ATP and an integral V_0 domain that translocated protons across the membranes. Structural model of the vacuolar V-ATPase showing subunit composition. The transmembrane domain (V_0) consists of the subunits a–d with several isoforms of the c subunit (denoted in small letters) and the cytosolic domain (V_1) made up of the A, B, C, D, E, F, G, and H subunits. Hydrolyzation of ATP is done on the intersection of the V_1A and V_1B subunits, and generated power rotate V-ATPase rotor formed by the V_0d, V_1D, and V_1F subunits. The "c-ring" couples the energy generation by ATP hydrolysis and translocation of the protons from the cytosol to the lumen.

Although V-ATPase subunits are highly conserved, some subunits have cell/tissue specific isoforms that govern V-ATPase subcellular localization. These isoforms are associated with subsets of V-ATPases that perform specialized functions. However, specialized V-ATPases represent a mixture of cell type selective isoforms and ubiquitous isoforms [13, 21–23]. Mammals have different isoforms for subunits V_0a, V_0d, V_1B, V_1C, V_1E, and V_1G, besides the ubiquitous ones. Subunit V_0a is most important in determining the subcellular localization of the V-ATPase, it has four isoform in humans which are found in different tissues and guide the subcellular locations of the V-ATPase. Isoform V_0a1 is present in the V-ATPase of the presynaptic plasma membrane and synaptic vesicles [24]. V_0a2 is found on the plasma membrane of the mammary epithelial cells [25]. V_0a2 is also found on the renal proximal tubule cells [26] and sperm acrosomes [27]. V_0a3 is found in the V-ATPase on the plasma membrane in the ruffled borders of the osteoclast [28], secretory endocrine tissues [29], pancreatic islets [22] and premature melanosomes [30, 31]. While V_0a4 is found in the renal intercalated cells [31] and clear cells of epididymis [32]. V_0a3 and V_0a4 isoforms are also overexpressed in tumor tissues, with V_0a4 primarily present on the plasma membrane and responsible for acidification of the extracellular matrix [33]. Other subunits in mammals that have multiple isoforms are V_1G, which has three and V_1B, V_1C, V_1E and V_0d all have two [34]. V_1B1 is expressed in renal and epididymal cells [35], while V_1E1 in testis and acrosome [27]. V_1C2 in the lungs and kidneys [36], V_1G3 in kidneys [36] and V_0d2 in kidneys and bones [37].

V-ATPase are responsible for acidification of endosomal, lysosomal compartments in the cell. In addition, they participate in other biological processes, such as toxin delivery, viral entry, membrane targeting, apoptosis, regulation of cytoplasmic pH, proteolytic process, and acidification of intracellular systems, are important roles of V-ATPases. Plasma membrane V-ATPase are responsible for the acidification of the urine in kidney and the FVreabsorption of bicarbonate ions. They help in bone resorption in osteoclast and facilitate tumor metastasis. Maintain the acidification of the sperm acrosome and activation of the different hydrolytic enzymes to ensure fertilization with the ovum.

2.2 V-ATPase regulation

Transmembrane proton transport by the V-ATPase is regulated in several different ways to modify pH in extracellular compartments or within intracellular vesicles. It is regulated by assembly process to form the holoenzyme and/or by trafficking to the appropriate cellular location.

2.2.1 Reversible assembly and disassembly

V-ATPase is a multi-subunit complex comprising of two distinct domains, the membrane integrated domain V_0, responsible for proton pumping and the free cytosolic domain V_1, which carries out the ATP-binding and hydrolysis. As the V_0 domain is membrane integrated, its subunits are polymerized on the rough endoplasmic reticulum during the translation process and are processed *via* the vesicular transport pathway. Whereas the subunits of the V_1 domain are synthesized on the free cytosolic ribosomes. Different subunits of each domain assemble to form the V_1 and V_0 domain before the formation of the V-ATPase holoenzyme. To form the complete functional V-ATPase holoenzyme the association of the V_0 and V_1 domain is essential. The efficiency of the V-ATPase function is dependent on its assembly process, as the

V_1 and V_0 domain independently are incapable of performing ATP hydrolysis and proton pumping, which are their respective functions [38, 39]. The phenomenon of the reversible assembly and disassembly of the V_1 and V_0 was first described for yeast, where it was noted that the V_1 domain dissociates from the V_0 domain in a reversible manner in the absence of glucose [20]. In one study with the *Manduca sexta* larval midgut the authors reported that the goblet cells of the apical membrane lose the proton pumping activity of the plasma membrane V-ATPase upon dissociation of its V_1 domain [40]. It is also known that the levels of cAMP and hormone induced protein kinase A (PKA) can regulate the plasma membrane V-ATPase assembly and activity in the blowfly salivary glands [41, 42]. It is shown that for the exocytosis of the presynaptic vesicles in the neuronal cells the V_1 disassembly is required on the mature exosomes, which fuse with the plasma membrane in the active zone and releases the exosomal contents in synaptic cleft [43]. The V_1 domain reassembles to V_0 on the neural membrane post release [43]. Using the cultured hamster kidney cells and sub cellular fractionation of different endosomal vesicles during maturation, it is known that the level of acidification in the lumen directly correlates with the level of V-ATPase of the isolated vesicles [4]. Glucose also affects the assembly of V-ATPase in human cells. Glucose starvation affects the V-ATPase assembly and activation by AMP kinase and phosphatidylinositide 3-kinase (PI3K)/Akt signaling pathway [44, 45]. The level of V-ATPase also goes high in lysosomes during dendritic cell maturation [46]. Although the mechanistic understanding of the assembly process is not well understood and needs more research to decipher it, however it is shown that subunit V_1C plays a central role in establishing interaction between the V_1 and V_0 domain during assembly assisted by their RAVE (regulator of the ATPase of vacuolar and endosomal membranes) complex in yeast [47–49].

2.2.2 Regulated trafficking of the V-ATPase

A second mechanism of controlling V-ATPase activity is *via* regulated trafficking of the functional holoenzyme. This occurs primarily in acid-secreting cells in a variety of different tissues [8, 50]. In proton-secreting intercalated cells of the kidney collecting duct and analogous organs in lower vertebrates and amphibian epidermis, regulation of transepithelial proton secretion is achieved by the exo- and endocytotic recycling of tubulovesicular structures containing high levels of V-ATPase holoenzymes in their membranes [8]. Trafficking of the V-ATPase to the plasma membrane and organelle membrane is also used by osteoclast [51] and the epididymal cells [52] to maintain the acidic pH of the extra cellular space and epididymal fluid respectively. The assembly and localization of V-ATPase are linked processes and are regulated by cellular needs. Although more studies are needed to establish the mechanism of assembly and trafficking process, but it is shown that the disassembly of V_1 from holocomplex is required for endosomal vesicles to localize the cargo to the plasma membrane and then V1 is reassembled to form the holoenzyme complex [43].

Apart from the signaling molecules mentioned above there are other kinases and proteases that are also involved in the assembly and trafficking of V-ATPase. In the intercalated renal cells G-protein coupled receptor Gpr116 is shown to negatively regulate the surface expression of proton pump V-ATPase [53]. Cytoskeletal proteins have a well-established role in trafficking the cargo from cytosol to the plasma membrane and vice versa. It has been shown that subunits V_1B and V_1C are associated with the actin of cytoskeleton and are essential for the movement of the V-ATPase cargo to the plasma membrane [54]. Profilin is a protein involved in actin polymerization. Subunits V_1B1 and V_1B2 also have a profilin-like domain [55]. Research has shown that use of microtubule depolymerizing drugs colchicine and vinblastine on turtles inhibits the excretion of protons in their urine upon carbon dioxide exposure, which alters the plasma pH [56]. Microtubule depolymerizing drugs also inhibit the V-ATPase localization and function in the renal intercalated cells [57] and epididymal clear cells [58]. PKA mediated phosphorylation of the subunits V_1A and V_1C upon increase in cAMP levels is necessary for the increase in the expression levels of V-ATPase on cell surface [12, 42, 59]. Activation of PKA upon bicarbonate stimulation is also essential for sensing acid-base balance and proton excretion by the kidneys [60, 61]. Furthermore, AMP kinase also phosphorylates the V-ATPase subunit V_1A and regulates its trafficking in renal epithelial cells [62]. The regulation of V-ATPase by phosphorylation is an interesting area for understanding the many different patterns of expression and regulation of V-ATPase activity in a variety of cells and tissues, as well as its pathophysiological dysfunction leading to human disease.

3. Physiological function of V-ATPase

3.1 Function of intracellular V-ATPases

The pH of cell and organelle lumen is an important governing parameter for the function of various organelles and is mainly controlled by V-ATPase-dependent proton transport. Receptor recycling and release of the ligands internalized *via* the receptor mediated endocytosis requires acidic pH of the endosomal lumen, which is maintained by the V-ATPase [63]. Density of the V-ATPase receptor on the cell surface is also synchronized by receptor recycling and it impacts the response and sensitivities for hormones and growth factors. In many cases ligand-receptor dissociation allows both protease delivery to lysosomes and the return of Mannose 6-phosphate receptors (MPRs) to the trans-Golgi network [64]. Acidification of endosomal lumen also plays important role in the formation of certain carrier vesicles, for transport of the cargo in the endocytic and secretory pathways [65]. When low pH is found within endosomes many pathogens take entry in cytoplasm. Low endosomal pH also promotes the entry of pathogenic agents such as diphtheria toxin and anthrax toxin, which first enter endosomes and then are released from late endosomes [66]. Acidic pH of the cytoplasm helps in fusion of enveloped viruses such as Influenza and Ebola, which is required for the insertion of viral genomes into the cytosol [67]. Some secretory vesicles are acidified by the V-ATPase to facilitate the proteolytic processing in prohormones such as proinsulin [68], in dendritic cell lysosomes [69], and in neurotransmitter antiporters [5, 70]. In lysosomes, a variety of proton/amino acid symporters use the proton gradient to drive amino acid efflux [71]. Lysosomal enzymes require acidic pH for activity, and for proper degradation of macromolecules [72]. These macromolecules are brought to lysosomes either endocytically *via* chaperone-mediated autophagy, or through macroautophagy, a catabolic program for recycling cellular components [73–75]. During autophagy process, acidification is essential for both autophagosomes and lysosome fusion as well as subsequent breakdown of luminal contents [76–78]. In normal condition, autophagy occurs at low basal levels but can be upregulated during times of energy stress or starvation (**Figure 2**) [79].

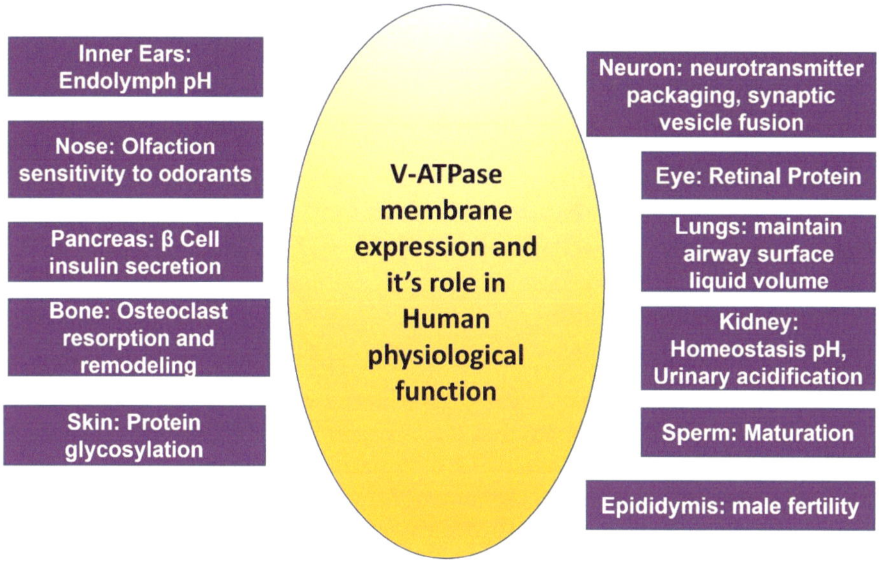

Figure 2.
The physiological importance of V-ATPase expression in membranes of different organs and tissues. Specific V-ATPase holoenzymes are expressed typically at the apical surface of proton-secreting cells in numerous tissues throughout the body and plays unique roles on place. Change in expression level of V-ATPase in specific tissues effected unique physiological role of these tissues/organs.

3.2 Function of plasma membrane V-ATPases

Renal α-intercalated cells, osteoclasts, cells of the epididymis, sustentacular cells of the olfactory epithelium and many polarized animal cell's plasma membrane have V-ATPases for transport of protons to the extracellular space [5, 52, 80]. Mutations in subunit V_0a3, of the plasma membrane V-ATPase of osteoclasts cause severe congenital form of osteopetrosis in humans [81, 82]. Renal α-intercalated cells respond to alterations in plasma pH by rapidly adjusting the density of apical V-ATPases to pump out the excess acid from the blood into the urine to be excreted out and restore the plasma pH. Studies have shown that distal renal tubular acidosis is associated with mutations in the plasma membrane V-VATPase subunit of the α-intercalated cells [31, 83]. Similarly clear cells of the epididymal epithelium regulate the acidic pH of the epididymal fluid to keep the spermatozoa in quiescent stage for storage and proper maturation [5, 52]. Loss of V-ATPase from the plasma membrane of epididymis results in increased epididymal fluid pH, defective sperms and renders the mice infertile [14]. The V-ATPase are also significant in cancer progression and metastasis [84]. **Figure 2** summarizes the broad localization and function of V-ATPases.

4. Emerging functions of the V-ATPase

4.1 Role in cancers

Recent studies revealed the role and significance of V-ATPase in cancer. It is shown that the plasma membrane V-ATPase help maintain an alkaline intracellular environment favorable for growth and an acidic extracellular environment favorable for invasion by proton efflux from the cell [85]. V-ATPase are shown to have higher expression in proliferating cancer cells of breast, prostate, lung, ovarian, liver, pancreatic, melanoma and esophageal cancers [10]. Increased expression of V-ATPase on the plasma membrane of the breast cancer cells correlates with increased invasiveness and metastatic potential of the breast cancer cell lines [86]. The increased metastatic potential is due to decreased pH of the extracellular matrix activating the proteases that degrade the extracellular matrix and aids in epithelial mesenchymal transition.

4.2 Immunomodulation

The V_0a2 isoform of Vacuolar ATPase has an immunomodulatory role in cancer and pregnancy. Research has shown that V_0a2 is required for normal sperm maturation and production in addition to embryo implantation [87, 88]. In the tumor microenvironment, the N terminal domain of V_0a2 polarizes monocytes to become tumor-associated macrophages (M2 type) and stimulates different monocyte subsets through the endocytosis pathway [89]. Studies demonstrated that V_0a2 deficiency in tumor cells alters the resident macrophage population in the tumor microenvironment and affects *in vivo* tumor growth [90]. Subunit V_0a2 is expressed on the primary granules of neutrophils and maintains the pH in exocytotic pathway for neutrophil activation [91]. These studies indicate V-ATPase importance as an immunomodulator in immune responses.

4.3 Warburg effect

Shifting of cancer cells from oxidative phosphorylation to aerobic glycolysis for energy production is referred as the Warburg effect [92]. Robust glycolytic cancer cells produce lots of acid and need an efficient proton pumping system to restore the intracellular pH homeostasis. Several studies have shown that for this purpose cancer cells rely on V-ATPase more than any other proton exchangers like Na^+H^+ exchangers, bicarbonate transporters and proton-lactate symporters to restore the alkaline intracellular pH [93]. V-ATPase also facilitate the activation of hypoxia induced factor 1 (HIF-1) in glycolytic cancer cells which promotes their growth [94].

4.4 Acid proteases

Dissolution of extracellular matrix is an essential process needed for the initiation of cancer invasion and metastasis. Proteases including cathepsins, metal requiring matrix metalloproteinases (MMP) and gelatinases carry out dissolution of extracellular matrix [95–97]. All these proteases are proenzymes that need an acidic pH for activation. The V-ATPase are involved in acidification of the extracellular space around the tumor to activate these proteinases and thus facilitate tumor invasion.

4.5 Drug resistance and V-ATPase inhibitors

Change in pH of microenvironment may influence sensitivity of tumor cells to chemotherapeutic drugs [98]. Recent studies suggests that the use of V-ATPase inhibitors not only causes cytosolic pH alterations leading to cell death but also enhances drug uptake, thereby making an effective component of combinatorial treatment to cancer [99]. In ovarian cancer, V_0a2 expression contributes in cisplatin mediated drug resistance and selective inhibition of V_0a2 could serve as an efficient strategy to treat chemo-resistant [100]. Currently, Apicularen and archazolids are reported to be potent and specific inhibitors of V-ATPase [101]. Thus combinatorial use of small molecule inhibitors for V-ATPase along with cancer drugs will be an effective strategy to treat/combat multi drug resistance cancers [102].

4.6 Autophagy

Autophagy is the natural process of selective degradation or recycling of macromolecules by autophagosomes to lysosomes [103]. In tumors, cells show dependency on autophagy as tumor progress from primary metastatic stage [104]. The proton pumping activity of V-ATPase is responsible for activation of lysosomal acid hydrolases, which degrade cargo uptake from autophagosomes [105]. Reports confirm the requirement of functional V-ATPase for autophagy [106]. Additionally V-ATPase inhibitor Bafilomycin is used as classic inhibitor of autophagy [107], but the exact role of V-ATPase in membrane dynamics of autophagy flux is not clear.

4.7 Signaling

The endo-lysosomal pathway is important for both positive and negative regulation of signaling pathways [108, 109]. The involvement of V-ATPase in signaling was first reported, by showing that inhibition of V-ATPase by Bafilomycin affected internalization of the epidermal growth factor receptor (EGFR) [77]. Studies demonstrated, V-ATPase has been also involved in multiple signal transduction pathways [110] like Notch, Wnt, transforming growth factor-β (TGF-β) and mammalian Target Of Rapamycin (m-TOR). Notch signaling depends on the endolysosomal pathway for its activation, maintenance and degradation of its key pathway mediators [111–113]. Some reports show that through its involvement in acidification of endolysosomal pathway, V-ATPase is required for the activation of Notch in endosomes as well as for its degradation in the lysosomes of Drosophila and mammalian cells [48, 114–117]. V-ATPase and Notch crosstalk is significantly important for normal growth as well as in Alzheimer's and cancer [118]. Wnt signaling pathway regulates numerous physiological processes. Dysregulation of Wnt pathway is linked to various pathologies including tumor metastasis [119–121]. The ATP6ap2 acts as an adaptor molecule between V-ATPase and Wnt receptor complex LRP 5/6 [122]. Furthermore, V-ATPase indirectly regulates Wnt signaling mediator β-catenin through Notch mediator NICD and autophagy [119, 123]. Mutations in V_0a2 are associate with elevated TGFβ signaling in patients with Cutis Laxa disease due to glycosylation defects [124]. V_0a2 inhibition activates Wnt signaling in a specific subtype of breast cancer called triple negative breast cancer (TNBC) and TGF-β pathway in mammary epithelial cells [25, 125]. mTOR regulates cellular growth during stress. Upon stimulation by amino acids during stress, V-ATPase activate the cascade of signaling events *via* RagA and RagC followed by GTP hydrolysis and loading of the mTOR-complex1 (mTORC1) to the lysosomal surface and activated mTORC1 switches the antigrowth to pro-growth signals [126–129].

5. V-ATPases in human disease

5.1 Cancer

As mentioned above the role of V-ATPase in cancer is evident. V-ATPases contribute to the survival and spread of cancer cells through several mechanisms. One of the ways that V-ATPases have been proposed to promote tumor cell survival is by maintaining an alkaline cytosolic pH, in contrast to normal cells which use the $Na^{+}K^{+}$ proton pump to maintain their pH. Tumor cells with hypoxia and high glycolytic metabolic stage have elevated levels of cytosolic acid [130]. Reports indicates that cancer cells increase V-ATPase biosynthesis and its targeting to the plasma membrane in order to secrete this increased proton extracellularly and restore the intracellular pH to support cell growth [131]. Studies have shown that V-ATPase is localized in plasma membrane of human breast tumors, lung tumors, osteosarcoma and numerous other cancer cell lines, including Ewing sarcoma, melanoma, breast, liver, pancreatic, prostate and ovarian cancer [33, 86, 99, 100, 132–137]. Blocking acid extrusion from the cancer cells after treating with V-ATPAse inhibitors has shown to increase apoptosis of these cells [138–141]. Decreased pH of the extracellular milieu driven by the V-ATPase of the cancer cells, can modify chemotherapeutic drugs by protonation [98], reduces drug uptake, its retention in the cytosol and cytotoxic effect on tumor [142, 143]. Thus, there is an enhanced efficacy of the chemotherapeutic drugs when used in combination with V-ATPase inhibitors [144, 145]. Some V-ATPase mediated mechanisms can be cancer subtype specific, as seen for prostate cancer. Prostate cancer cells need androgen receptor for proliferation. Hypoxia-inducible factor 1-alpha (HIF1α) is a transcriptional repressor for androgen receptors [146, 147]. A recent study showed that inhibition of V-ATPase, reduces prostate cancer growth by reducing the iron-dependent hydroxylation followed by degradation of HIF1α [146]. Overexpression of cathepsins, is associated with worse prognosis for different human cancers [148]. Inhibition of cathepsins reduces metastasis and spread of breast cancer in mice [149, 150]. Activation of secreted cathepsins happens in the acidic extracellular space, which are acidified by the plasma membrane V-ATPase. V-ATPases have been detected at the plasma membrane of numerous invasive cancer cell lines. Since plasma membrane targeting is controlled by isoforms of subunit a, it is likely that cancer cells will upregulate particular isoforms in order to increase localization of V-ATPases to the plasma membrane [33, 151]. Immunofluorescence studies in breast carcinoma showed that the levels of V_0a3 isoform are higher in the invasive tumor cells relative to non-invasive and normal breast tissues [132], and inhibition of V_0a3 reduces metastasis of murine melanoma [137]. Study with prostate cancer cell line PC3, demonstrate that there is an increased expression of V_0a1 and V_0a3 isoforms on the plasma membrane and siRNA mediated knock down of these isoforms reduces it growth and in-invasion in cell culture [136]. While V_0a2 is expressed in ovarian cancer cell lines [100], V_0a4 is shown to be overexpressed in metastatic breast cancer cell line MDA-MB231 [33]. Inhibition of V-ATPase hinders the activity of matrix metalloproteinase (MMP) MMP2 and MMP9 in different cancers *in vitro* [100, 135] and *in vivo* [137]. Significance of the V_0a4 isoform in invasion and metastasis of breast tumors is established by the CRISPR/ Cas9 mediated knock down of the *ATP6V0A4* gene in the murine breast cancer cell line 4 T1 and loss of its metastatic ability [152]. Different subunits have significance for specific cancer's, V_1G1 is necessary for stem cell in neurospheres [153], V_1E1 in pancreatic cancer cells [135] and V_1A1 for gastric cancer [154].

5.2 Osteoporosis and Osteopetrosis

Healthy bone mass contributes to a healthy skeleton, which is based on the synchronized activity of the osteoblasts (the bone forming cells) and osteoclast (the cells for dissolution and reabsorption of the bone matrix). Osteoclasts are multinucleated cells that attaches to the bone surface with their ruffled borders and create a very acidic compartment called resorption lacunae. It is in the resorption lacunae that solubilization and degradation of the extracellular matrix, collagen fibers and the bone matrix happens. V-ATPase are located in the ruffle borders of the osteoclast and are responsible for the maintaining the acidity of the resorption lacunae. The acidic pH of the resorption lacunae is important for activation of the multiple hydrolases needed for bone dissolution. Lack of osteoclast functioning can cause increased bone density, diminished bone strength and several skeletal defects, a condition referred as osteopetrosis. V_0a3 is overexpressed in the highly resorptive osteoclast [155]. V_1C1 is also present with V_0a3 in the ruffle borders of the osteoclast [156]. As shown by RNAi studies the isoforms V_0a3 and V_1C1 are essential for the acidic pH of the resorption lacunae [156]. Isoform V_0d2 is needed for cell fusion during osteoclast maturation [157] and V_1B2 is also expressed in the ruffle borders [158]. Mutations in the gene *TCIRG1* that encodes V_0a3 are the primary cause of infantile malignant autosomal recessive osteopetrosis (present in about 50% of the cases), a rare congenital disease caused by the failure of osteoclasts function.

Another disorder associated with V-ATPase in osteoclast is osteoporosis, which is characterized by low bone mineral density due to increased bone degradation by osteoclast and low bone formation by osteoblast. Single nucleotide polymorphism in the *ATP6V1G1* gene, which encodes V_1G1 and total or partial loss of function of V_1H subunit are shown to have association with osteoporosis [159–161].

5.3 Neurodegenerative diseases

Mutations in V-ATPase subunits isoforms are cause of different neurodegenerative disorders. Autophagy is a housekeeping process involved in the removal of abnormal and misfolded proteins and damaged organelles from the cells. Autophagy is dependent on the lysosomal function, which are heavily dependent on the acidic pH of the lysosomal compartments maintained by the V-ATPases. Autophagy is very important for terminally differentiated neuronal cells as shown by neurodegeneration in the mice upon inhibition of autophagic process [162, 163]. Many neurodegenerative disorders including Alzheimer's disease (AD) are characterized by pathological hallmark, like increase in the misfolded protein aggregates in the brain. AD is characterized by the extracellular plagues made up of insoluble amyloid β (Aβ) fibers [164]. Proteases α, β and γ-secretase are needed for proper processing of amyloid protein. Presenilin-1 (PS1) is a cofactor for γ-secretase and mutations in *PS1* gene are associated with familial AD [165]. PS1 is also needed for accurate subcellular trafficking of V_0a1 to the neuronal lysosomes. Mutation in PS1 also affect V_0a1 trafficking and lysosomal acidification, rescue of lysosomal acidification in PS1 knock out cells reduces Aβ build up [118]. Parkinsons, another neurodegenerative disorder is caused by buildup of α-synuclein aggregates [166]. The *ATP6AP2* gene encodes the accessory subunit for V-ATPase. Mutation in *ATP6AP2*, also known as Renin/Prorenin receptor, causes X-linked Parkinsonism with spasticity, an early-onset form of Parkinsonism with defective lysosomal acidification [167]. Mutation in *ATP6AP2* is also associated with the Wolfram syndrome, a neurodegenerative disorder characterized by endoplasmic reticulum stress, childhood diabetes mellitus, severe neurological disabilities [168] and X-linked mental retardation Hedera type (MRXSH), a congenital disorder of intellectual disability, delayed motor and speech development and epilepsy [169, 170]. Many other neurodegenerative disorders which are not directly caused by V-ATPase defects nevertheless exhibit lysosomal impairment [171]. Restoration of lysosomal function therefore represents an attractive therapeutic concept that should be investigated further.

5.4 Distal renal tubular acidosis (DRTA) and hearing loss

Intercalated cells of the kidney are the primary regulators of the physiological urine acidification. They sense the physiological changes in the acidosis/alkalosis levels and balance it by reorganize the V-ATPase on the apical membrane. The V-ATPase on the plasma membrane is responsible for acidification of the urine and maintainence of the physiological pH by the kidneys [23, 80, 172]. The isoforms V_1B1 and V_0a4 are characteristics of the apical membrane V-ATPase [173, 174]. As these V-ATPase are needed for urine acidification, mutation in the genes *ATP6V1B1* and *ATP6V0A4* encoding the renal isoforms leads to an autosomal recessive genetic disease called DRTA [83, 175]. It has been shown that the mutation in the genes causes V0a4 retention in the endoplasmic reticulum and not being able to perform the protonation of the urine is the cause for DRTA onset [176]. It has also been shown that mice lacking V_1B1 and V_0a4 develop symptoms similar to human DRTA like metabolic acidosis, hypokalemia and hearing loss [177–179]. This is treated by administering bicarbonates to regulate metabolic acidosis. Patients with *ATP6V0A4* mutation also present with hearing loss. V_1B1 and V_0a4 isoforms are also expressed in the human inner ear and maintain the endolymph pH homeostasis, necessary for mechano transduction sensitivity and auditory function. V_0a4 knockout mice present severe deafness associated with enlarged cochlear and endolymphatic compartments [178]. Alkali treatment does not help restore the deafness; the conition is treated by the use of hearing devices.

5.5 Cutis laxa (CL) and wrinkly skin syndrome (WSS)

Cutis laxa is a skin condition, characterized by loss of elasticity in the skin tissue. Skin losses its strength to stretch and instead hangs in loose folds, becomes saggy and gives wrinkled appearance to the face and others parts of the body. CL is also associated with variable neurological and skeletal alterations. CL can be both inherited and/or acquired and is caused by autosomal recessive inheritance of V-ATPase subunit mutations. It is characterized by impaired Golgi function, glycosylation defects and delayed retrograde transport from Golgi to endoplasmic reticulum, thus resulting in abnormal elastic fibers that affect the skin and internal organs [180, 181]. WSS is also a type of CL caused by mutation in *ATP6V0A2* gene. Besides V_0a2 mutations, mutations in *ATP6V1E1* and *ATP6V1A*, were also recently associated with CL. However, each mutation result in CL, but they vary in clinical manifestation, as it is multisystemic and also includes risk of cardiopulmonary problems [181].

5.6 Other roles of V-ATPase

Subunit V_0a4 also targets the V-ATPase to the apical membrane of epididymal clear cells, but its association with male fertility are not well understood for patients with

V_0a4 mutations [8]. Some studies have shown that the levels of the V_0a2 is higher in the fertile male compared to infertile. Study also shows that the higher levels of V_0a2 are associated with Sperm capacitation [88]. Zimmermann-Laband syndrome (ZLS) is a rare genetic disorder characterized by gingival fibromatosis (abnormally large gum), defects in craniofacial features, nails, ear and nose. In some cases, ZLS is also associated with mental retardation. Two patient suffering with ZLS showed mutation in the *ATP6V1B2* gene resulting in substitution of proline instead of arginine in the V_1B2 subunit.

Viruses like influenzas virus [182], Sindbis virus [183],and West Nile virus [184] use the endosomal route for infecting the host cells and delivering its genetic material. Host V-ATPase are needed for endosomal compartment acidification, which also facilitates the uncoating of the virus and release of its genetic material. Pathogenic fungi use its V-ATPase in establishing the infection, as demonstrated that impairment of V-ATPase activity, either by V-ATPase inhibitors or deletion of specific subunits/assembly factors, dramatically diminishes or inhibits virulence-associated traits [185, 186]. For example, knocking down *VPH2* in *Candida albicans* reduced candidiasis in mice model [185]. Other examples are the V_0c subunit for *C. albicans* [186], V_0a for *Cryptococcus neoformans* [187] that affects fungal infectivity and thus the fungal V-ATPase subunits can be used as therapeutic targets.

6. Conclusions

V-ATPase are proton pumping ATPase with a housekeeping role of maintaining the pH of the cytosol, organelle lumen and extracellular space. V-ATPase is a multi-subunit complex with highly regulated assembly and trafficking to the right compartment. Its multi-subunit complex and different isoforms are the basis for its diverse location. It works by the rotary mechanism, has two domains: one membrane embedded, responsible for proton transport and other cytosolic, which carries out ATP hydrolysis. The reduced pH is in turn, required for processes that involve the trafficking of intracellular vesicles to their correct destination, post-translational modification of proteins in cellular compartments and the plasma membrane and activation of different proteases. It is needed for lysosomal function, autophagy, immunomodulation and endosomal maturation. V-ATPase are a key component of the renal apical layer and assist in maintaining the physiological pH and preventing metabolic acidosis. They are essential for osteoclast functioning which provides proper skeletal health by working in symphony with osteoblast. Increased plasma membrane activity of the V-ATPase is the reason for cancer metastasis. V-ATPase are also required for giving the proper skin texture. As discuss above, the V-ATPase is clearly involved in many aspects of normal physiological function, and mutation in the gene for different subunits either leading to lack of proper protein-protein interaction and/or assembly, mis-localization, loss of function of the subunits, or hyperactivity are attribute to different human diseases. There are lot of therapeutic opportunities for V-ATPases-directed therapies. Using inhibitors for the plasma membrane V-ATPase for cancer and osteoclast is a promising strategy for treating cancer metastasis and osteoporosis. Restoring the intracellular V-ATPase function could be a good approach for helping the neurodegenerative disorders associated with loss or reduced autophagy. Additionally targeting the endosomal V-ATPase can help reduce viral infection. Combating V-ATPase of the fungal pathogen can be an effective strategy to use as an antifungal drug. Although V-ATPases are known to play a role in sperm maturation and fertilization, their association to male fertility needs more research. Most of the treatment option for the V-ATPase mediated diseases are focused on elevating the symptoms not focused on eliminating the root cause. Thus, research is needed to focus on ways to rescue the activity of these disease associated mutants. Finding effective inhibitors for V-ATPase has been challenging due to their ubiquitous role, so far developed V-ATPase inhibitors are toxic and have off target effects. Thus rigorous research is needed to find effective inhibitors as increasing evidence is building, highlighting the role of V-ATPase in different human diseases.

Author details

Anuj Tripathi[1] and Smita Misra[2,3]*

1 Department of Microbiology, Immunology, and Physiology, Meharry Medical College, Nashville, TN, USA

2 School of Graduate Studies and Research, Meharry Medical College, Nashville, TN, USA

3 Centre for Women's Health Research (CWHR), Meharry Medical College, Nashville, TN, USA

*Address all correspondence to: smisra@mmc.edu

References

[1] Song Q, Meng B, Xu H, Mao Z. The emerging roles of vacuolar-type ATPase-dependent lysosomal acidification in neurodegenerative diseases. Translational Neurodegeneration. 2020;**9**:17

[2] Pamarthy S, Kulshrestha A, Katara GK, Beaman KD. The curious case of vacuolar ATPase: Regulation of signaling pathways. Molecular Cancer. 2018;**17**:41

[3] Nelson N, Perzov N, Cohen A, Hagai K, Padler V, Nelson H. The cellular biology of proton-motive force generation by V-ATPases. The Journal of Experimental Biology. 2000;**203**:89-95

[4] Lafourcade C, Sobo K, Kieffer-Jaquinod S, Garin J, van der Goot FG. Regulation of the V-ATPase along the endocytic pathway occurs through reversible subunit association and membrane localization. PLoS One. 2008;**3**:e2758

[5] Forgac M. Vacuolar ATPases: Rotary proton pumps in physiology and pathophysiology. Nature Reviews. Molecular Cell Biology. 2007;**8**:917-929

[6] Nelson N. Structural conservation and functional diversity of V-ATPases. Journal of Bioenergetics and Biomembranes. 1992;**24**:407-414

[7] Nelson N, Harvey WR. Vacuolar and plasma membrane proton-adenosinetriphosphatases. Physiological Reviews. 1999;**79**:361-385

[8] Breton S, Brown D. Regulation of luminal acidification by the V-ATPase. Physiology (Bethesda, Md.). 2013;**28**:318-329

[9] Marshansky V, Rubinstein JL, Gruber G. Eukaryotic V-ATPase: Novel structural findings and functional insights. Biochimica et Biophysica Acta. 2014;**1837**:857-879

[10] Cotter K, Stransky L, McGuire C, Forgac M. Recent insights into the structure, regulation, and function of the V-ATPases. Trends in Biochemical Sciences. 2015;**40**:611-622

[11] Maxson ME, Grinstein S. The vacuolar-type H(+)-ATPase at a glance—More than a proton pump. Journal of Cell Science. 2014;**127**:4987-4993

[12] Alzamora R, Thali RF, Gong F, Smolak C, Li H, Baty CJ, et al. PKA regulates vacuolar H+-ATPase localization and activity via direct phosphorylation of the a subunit in kidney cells. The Journal of Biological Chemistry. 2010;**285**:24676-24685

[13] Barvencik F, Kurth I, Koehne T, Stauber T, Zustin J, Tsiakas K, et al. CLCN7 and TCIRG1 mutations differentially affect bone matrix mineralization in osteopetrotic individuals. Journal of Bone and Mineral Research. 2014;**29**:982-991

[14] Blomqvist SR, Vidarsson H, Soder O, Enerback S. Epididymal expression of the forkhead transcription factor Foxi1 is required for male fertility. The EMBO Journal. 2006;**25**:4131-4141

[15] Brown D, Smith PJ, Breton S. Role of V-ATPase-rich cells in acidification of the male reproductive tract. The Journal of Experimental Biology. 1997;**200**:257-262

[16] Paunescu TG, Jones AC, Tyszkowski R, Brown D. V-ATPase expression in the mouse olfactory epithelium. American Journal

of Physiology. Cell Physiology. 2008;**295**:C923-C930

[17] Nishi T, Forgac M. The vacuolar (H+)-ATPases—Nature's most versatile proton pumps. Nature Reviews. Molecular Cell Biology. 2002;**3**:94-103

[18] Shao E, Forgac M. Involvement of the nonhomologous region of subunit A of the yeast V-ATPase in coupling and in vivo dissociation. The Journal of Biological Chemistry. 2004;**279**:48663-48670

[19] Wilkens S, Inoue T, Forgac M. Three-dimensional structure of the vacuolar ATPase. Localization of subunit H by difference imaging and chemical cross-linking. The Journal of Biological Chemistry. 2004;**279**:41942-41949

[20] Kane PM. Regulation of V-ATPases by reversible disassembly. FEBS Letters. 2000;**469**:137-141

[21] Serrano EM, Ricofort RD, Zuo J, Ochotny N, Manolson MF, Holliday LS. Regulation of vacuolar H(+)-ATPase in microglia by RANKL. Biochemical and Biophysical Research Communications. 2009;**389**:193-197

[22] Sun-Wada GH, Toyomura T, Murata Y, Yamamoto A, Futai M, Wada Y. The a3 isoform of V-ATPase regulates insulin secretion from pancreatic beta-cells. Journal of Cell Science. 2006;**119**:4531-4540

[23] Stehberger PA, Schulz N, Finberg KE, Karet FE, Giebisch G, Lifton RP, et al. Localization and regulation of the ATP6V0A4 (a4) vacuolar H+-ATPase subunit defective in an inherited form of distal renal tubular acidosis. Journal of the American Society of Nephrology. 2003;**14**:3027-3038

[24] Morel N, Dedieu JC, Philippe JM. Specific sorting of the a1 isoform of the V-H+ATPase a subunit to nerve terminals where it associates with both synaptic vesicles and the presynaptic plasma membrane. Journal of Cell Science. 2003;**116**:4751-4762

[25] Pamarthy S, Mao L, Katara GK, Fleetwood S, Kulshreshta A, Gilman-Sachs A, et al. The V-ATPase a2 isoform controls mammary gland development through notch and TGF-beta signaling. Cell Death & Disease. 2016;**7**:e2443

[26] Hurtado-Lorenzo A, Skinner M, El Annan J, Futai M, Sun-Wada GH, Bourgoin S, et al. V-ATPase interacts with ARNO and Arf6 in early endosomes and regulates the protein degradative pathway. Nature Cell Biology. 2006;**8**:124-136

[27] Sun-Wada GH, Imai-Senga Y, Yamamoto A, Murata Y, Hirata T, Wada Y, et al. A proton pump ATPase with testis-specific E1-subunit isoform required for acrosome acidification. The Journal of Biological Chemistry. 2002;**277**:18098-18105

[28] Toyomura T, Murata Y, Yamamoto A, Oka T, Sun-Wada GH, Wada Y, et al. From lysosomes to the plasma membrane: Localization of vacuolar-type H+ -ATPase with the a3 isoform during osteoclast differentiation. The Journal of Biological Chemistry. 2003;**278**:22023-22030

[29] Sun-Wada GH, Tabata H, Kawamura N, Futai M, Wada Y. Differential expression of a subunit isoforms of the vacuolar-type proton pump ATPase in mouse endocrine tissues. Cell and Tissue Research. 2007;**329**:239-248

[30] Tabata H, Kawamura N, Sun-Wada GH, Wada Y. Vacuolar-type H(+)-ATPase with the a3 isoform

is the proton pump on premature melanosomes. Cell and Tissue Research. 2008;**332**:447-460

[31] Oka T, Murata Y, Namba M, Yoshimizu T, Toyomura T, Yamamoto A, et al. a4, a unique kidney-specific isoform of mouse vacuolar H+-ATPase subunit a. The Journal of Biological Chemistry. 2001;**276**:40050-40054

[32] Pietrement C, Sun-Wada GH, Silva ND, McKee M, Marshansky V, Brown D, et al. Distinct expression patterns of different subunit isoforms of the V-ATPase in the rat epididymis. Biology of Reproduction. 2006;**74**:185-194

[33] Hinton A, Sennoune SR, Bond S, Fang M, Reuveni M, Sahagian GG, et al. Function of a subunit isoforms of the V-ATPase in pH homeostasis and in vitro invasion of MDA-MB231 human breast cancer cells. The Journal of Biological Chemistry. 2009;**284**:16400-16408

[34] Toei M, Saum R, Forgac M. Regulation and isoform function of the V-ATPases. Biochemistry. 2010;**49**:4715-4723

[35] Paunescu TG, Da Silva N, Marshansky V, McKee M, Breton S, Brown D. Expression of the 56-kDa B2 subunit isoform of the vacuolar H(+)-ATPase in proton-secreting cells of the kidney and epididymis. American Journal of Physiology. Cell Physiology. 2004;**287**:C149-C162

[36] Sun-Wada GH, Yoshimizu T, Imai-Senga Y, Wada Y, Futai M. Diversity of mouse proton-translocating ATPase: Presence of multiple isoforms of the C, d and G subunits. Gene. 2003;**302**:147-153

[37] Smith AN, Jouret F, Bord S, Borthwick KJ, Al-Lamki RS, Wagner CA, et al. Vacuolar H+-ATPase d2 subunit: Molecular characterization, developmental regulation, and localization to specialized proton pumps in kidney and bone. Journal of the American Society of Nephrology. 2005;**16**:1245-1256

[38] Sharma S, Oot RA, Khan MM, Wilkens S. Functional reconstitution of vacuolar H(+)-ATPase from Vo proton channel and mutant V1-ATPase provides insight into the mechanism of reversible disassembly. The Journal of Biological Chemistry. 2019;**294**:6439-6449

[39] Parra KJ, Keenan KL, Kane PM. The H subunit (Vma13p) of the yeast V-ATPase inhibits the ATPase activity of cytosolic V1 complexes. The Journal of Biological Chemistry. 2000;**275**:21761-21767

[40] Sumner JP, Dow JA, Earley FG, Klein U, Jager D, Wieczorek H. Regulation of plasma membrane V-ATPase activity by dissociation of peripheral subunits. The Journal of Biological Chemistry. 1995;**270**:5649-5653

[41] Dames P, Zimmermann B, Schmidt R, Rein J, Voss M, Schewe B, et al. cAMP regulates plasma membrane vacuolar-type H+-ATPase assembly and activity in blowfly salivary glands. Proceedings of the National Academy of Sciences of the United States of America. 2006;**103**:3926-3931

[42] Rein J, Voss M, Blenau W, Walz B, Baumann O. Hormone-induced assembly and activation of V-ATPase in blowfly salivary glands is mediated by protein kinase a. American Journal of Physiology. Cell Physiology. 2008;**294**:C56-C65

[43] Bodzeta A, Kahms M, Klingauf J. The presynaptic v-ATPase reversibly disassembles and thereby modulates exocytosis but is not part of the

fusion machinery. Cell Reports. 2017;**20**:1348-1359

[44] McGuire CM, Forgac M. Glucose starvation increases V-ATPase assembly and activity in mammalian cells through AMP kinase and phosphatidylinositide 3-kinase/Akt signaling. The Journal of Biological Chemistry. 2018;**293**:9113-9123

[45] Sautin YY, Lu M, Gaugler A, Zhang L, Gluck SL. Phosphatidylinositol 3-kinase-mediated effects of glucose on vacuolar H+-ATPase assembly, translocation, and acidification of intracellular compartments in renal epithelial cells. Molecular and Cellular Biology. 2005;**25**:575-589

[46] Trombetta ES, Ebersold M, Garrett W, Pypaert M, Mellman I. Activation of lysosomal function during dendritic cell maturation. Science. 2003;**299**:1400-1403

[47] Seol JH, Shevchenko A, Shevchenko A, Deshaies RJ. Skp1 forms multiple protein complexes, including RAVE, a regulator of V-ATPase assembly. Nature Cell Biology. 2001;**3**:384-391

[48] Sethi N, Yan Y, Quek D, Schupbach T, Kang Y. Rabconnectin-3 is a functional regulator of mammalian notch signaling. The Journal of Biological Chemistry. 2010;**285**:34757-34764

[49] Merkulova M, Paunescu TG, Azroyan A, Marshansky V, Breton S, Brown D. Mapping the H(+) (V)-ATPase interactome: Identification of proteins involved in trafficking, folding, assembly and phosphorylation. Scientific Reports. 2015;**5**:14827

[50] Wagner CA, Finberg KE, Breton S, Marshansky V, Brown D, Geibel JP. Renal vacuolar H+-ATPase. Physiological Reviews. 2004;**84**:1263-1314

[51] Matsumoto N, Sekiya M, Tohyama K, Ishiyama-Matsuura E, Sun-Wada GH, Wada Y, et al. Essential role of the a3 isoform of V-ATPase in secretory lysosome trafficking via Rab7 recruitment. Scientific Reports. 2018;**8**:6701

[52] Shum WW, Da Silva N, Brown D, Breton S. Regulation of luminal acidification in the male reproductive tract via cell-cell crosstalk. The Journal of Experimental Biology. 2009;**212**:1753-1761

[53] Zaidman NA, Tomilin VN, Hassanzadeh Khayyat N, Damarla M, Tidmore J, Capen DE, et al. Adhesion-GPCR Gpr116 (ADGRF5) expression inhibits renal acid secretion. Proceedings of the National Academy of Sciences of the United States of America. 2020;**117**:26470-26481

[54] Vitavska O, Wieczorek H, Merzendorfer H. A novel role for subunit C in mediating binding of the H+-V-ATPase to the actin cytoskeleton. The Journal of Biological Chemistry. 2003;**278**:18499-18505

[55] Chen SH, Bubb MR, Yarmola EG, Zuo J, Jiang J, Lee BS, et al. Vacuolar H+-ATPase binding to microfilaments: Regulation in response to phosphatidylinositol 3-kinase activity and detailed characterization of the actin-binding site in subunit B. The Journal of Biological Chemistry. 2004;**279**:7988-7998

[56] Arruda JA, Sabatini S, Mola R, Dytko G. Inhibition of H+ secretion in the turtle bladder by colchicine and vinblastine. The Journal of Laboratory and Clinical Medicine. 1980;**96**:450-459

[57] Brown D, Sabolic I, Gluck S. Colchicine-induced redistribution of proton pumps in kidney epithelial cells.

Kidney International. Supplement. 1991;**33**:S79-S83

[58] Breton S, Nsumu NN, Galli T, Sabolic I, Smith PJ, Brown D. Tetanus toxin-mediated cleavage of cellubrevin inhibits proton secretion in the male reproductive tract. American Journal of Physiology. Renal Physiology. 2000;**278**:F717-F725

[59] Voss M, Vitavska O, Walz B, Wieczorek H, Baumann O. Stimulus-induced phosphorylation of vacuolar H(+)-ATPase by protein kinase a. The Journal of Biological Chemistry. 2007;**282**:33735-33742

[60] Brown D, Wagner CA. Molecular mechanisms of acid-base sensing by the kidney. Journal of the American Society of Nephrology. 2012;**23**:774-780

[61] Levin LR, Buck J. Physiological roles of acid-base sensors. Annual Review of Physiology. 2015;77:347-362

[62] Alzamora R, Al-Bataineh MM, Liu W, Gong F, Li H, Thali RF, et al. AMP-activated protein kinase regulates the vacuolar H+-ATPase via direct phosphorylation of the A subunit (ATP6V1A) in the kidney. American Journal of Physiology. Renal Physiology. 2013;**305**:F943-F956

[63] Maxfield FR, McGraw TE. Endocytic recycling. Nature Reviews. Molecular Cell Biology. 2004;**5**:121-132

[64] Ghosh P, Dahms NM, Kornfeld S. Mannose 6-phosphate receptors: New twists in the tale. Nature Reviews. Molecular Cell Biology. 2003;**4**:202-212

[65] Gu F, Gruenberg J. ARF1 regulates pH-dependent COP functions in the early endocytic pathway. The Journal of Biological Chemistry. 2000;**275**:8154-8160

[66] Gruenberg J, van der Goot FG. Mechanisms of pathogen entry through the endosomal compartments. Nature Reviews. Molecular Cell Biology. 2006;**7**:495-504

[67] Grove J, Marsh M. The cell biology of receptor-mediated virus entry. The Journal of Cell Biology. 2011;**195**:1071-1082

[68] Rhodes CJ, Lucas CA, Mutkoski RL, Orci L, Halban PA. Stimulation by ATP of proinsulin to insulin conversion in isolated rat pancreatic islet secretory granules. Association with the ATP-dependent proton pump. The Journal of Biological Chemistry. 1987;**262**:10712-10717

[69] Trombetta ES, Mellman I. Cell biology of antigen processing in vitro and in vivo. Annual Review of Immunology. 2005;**23**:975-1028

[70] Farsi Z, Preobraschenski J, van den Bogaart G, Riedel D, Jahn R, Woehler A. Single-vesicle imaging reveals different transport mechanisms between glutamatergic and GABAergic vesicles. Science. 2016;**351**:981-984

[71] Xu H, Ren D. Lysosomal physiology. Annual Review of Physiology. 2015;**77**:57-80

[72] Mindell JA. Lysosomal acidification mechanisms. Annual Review of Physiology. 2012;**74**:69-86

[73] Nakatogawa H, Suzuki K, Kamada Y, Ohsumi Y. Dynamics and diversity in autophagy mechanisms: Lessons from yeast. Nature Reviews. Molecular Cell Biology. 2009;**10**:458-467

[74] Dice JF. Chaperone-mediated autophagy. Autophagy. 2007;**3**:295-299

[75] Galluzzi L, Bravo-San Pedro JM, Levine B, Green DR, Kroemer G.

Pharmacological modulation of autophagy: Therapeutic potential and persisting obstacles. Nature Reviews. Drug Discovery. 2017;**16**:487-511

[76] Feng Y, He D, Yao Z, Klionsky DJ. The machinery of macroautophagy. Cell Research. 2014;**24**:24-41

[77] Yoshimori T, Yamamoto A, Moriyama Y, Futai M, Tashiro Y. Bafilomycin A1, a specific inhibitor of vacuolar-type H(+)-ATPase, inhibits acidification and protein degradation in lysosomes of cultured cells. The Journal of Biological Chemistry. 1991;**266**:17707-17712

[78] Kawai A, Uchiyama H, Takano S, Nakamura N, Ohkuma S. Autophagosome-lysosome fusion depends on the pH in acidic compartments in CHO cells. Autophagy. 2007;**3**:154-157

[79] Mizushima N, Komatsu M. Autophagy: Renovation of cells and tissues. Cell. 2011;**147**:728-741

[80] Brown D, Paunescu TG, Breton S, Marshansky V. Regulation of the V-ATPase in kidney epithelial cells: Dual role in acid-base homeostasis and vesicle trafficking. The Journal of Experimental Biology. 2009;**212**:1762-1772

[81] Toyomura T, Oka T, Yamaguchi C, Wada Y, Futai M. Three subunit a isoforms of mouse vacuolar H(+)-ATPase. Preferential expression of the a3 isoform during osteoclast differentiation. The Journal of Biological Chemistry. 2000;**275**:8760-8765

[82] Frattini A, Orchard PJ, Sobacchi C, Giliani S, Abinun M, Mattsson JP, et al. Defects in TCIRG1 subunit of the vacuolar proton pump are responsible for a subset of human autosomal recessive osteopetrosis. Nature Genetics. 2000;**25**:343-346

[83] Smith AN, Skaug J, Choate KA, Nayir A, Bakkaloglu A, Ozen S, et al. Mutations in ATP6N1B, encoding a new kidney vacuolar proton pump 116-kD subunit, cause recessive distal renal tubular acidosis with preserved hearing. Nature Genetics. 2000;**26**:71-75

[84] Stransky L, Cotter K, Forgac M. The function of V-ATPases in cancer. Physiological Reviews. 2016;**96**:1071-1091

[85] Sennoune SR, Martinez-Zaguilan R. Plasmalemmal vacuolar H+-ATPases in angiogenesis, diabetes and cancer. Journal of Bioenergetics and Biomembranes. 2007;**39**:427-433

[86] Sennoune SR, Bakunts K, Martinez GM, Chua-Tuan JL, Kebir Y, Attaya MN, et al. Vacuolar H+-ATPase in human breast cancer cells with distinct metastatic potential: Distribution and functional activity. American Journal of Physiology. Cell Physiology. 2004;**286**:C1443-C1452

[87] Jaiswal MK, Agrawal V, Katara GK, Pamarthy S, Kulshrestha A, Chaouat G, et al. Male fertility and apoptosis in normal spermatogenesis are regulated by vacuolar-ATPase isoform a2. Journal of Reproductive Immunology. 2015;**112**:38-45

[88] Jaiswal MK, Mallers TM, Larsen B, Kwak-Kim J, Chaouat G, Gilman-Sachs A, et al. V-ATPase upregulation during early pregnancy: A possible link to establishment of an inflammatory response during preimplantation period of pregnancy. Reproduction. 2012;**143**:713-725

[89] Kwong C, Gilman-Sachs A, Beaman K. Tumor-associated a2 vacuolar ATPase acts as a key mediator of cancer-related inflammation by inducing pro-tumorigenic properties

in monocytes. Journal of Immunology. 2011;**186**:1781-1789

[90] Katara GK, Kulshrestha A, Jaiswal MK, Pamarthy S, Gilman-Sachs A, Beaman KD. Inhibition of vacuolar ATPase subunit in tumor cells delays tumor growth by decreasing the essential macrophage population in the tumor microenvironment. Oncogene. 2016;**35**:1058-1065

[91] Gilman-Sachs A, Tikoo A, Akman-Anderson L, Jaiswal M, Ntrivalas E, Beaman K. Expression and role of a2 vacuolar-ATPase (a2V) in trafficking of human neutrophil granules and exocytosis. Journal of Leukocyte Biology. 2015;**97**:1121-1131

[92] Liberti MV, Locasale JW. The Warburg effect: How does it benefit cancer cells? Trends in Biochemical Sciences. 2016;**41**:211-218

[93] Torigoe T, Izumi H, Ise T, Murakami T, Uramoto H, Ishiguchi H, et al. Vacuolar H(+)-ATPase: Functional mechanisms and potential as a target for cancer chemotherapy. Anti-Cancer Drugs. 2002;**13**:237-243

[94] Gillies RJ, Robey I, Gatenby RA. Causes and consequences of increased glucose metabolism of cancers. Journal of Nuclear Medicine. 2008;**49**(Suppl. 2): 24S-42S

[95] Kubisch R, Frohlich T, Arnold GJ, Schreiner L, von Schwarzenberg K, Roidl A, et al. V-ATPase inhibition by archazolid leads to lysosomal dysfunction resulting in impaired cathepsin B activation in vivo. International Journal of Cancer. 2014;**134**:2478-2488

[96] Kubota S, Seyama Y. Overexpression of vacuolar ATPase 16-kDa subunit in 10T1/2 fibroblasts enhances invasion with concomitant induction of matrix metalloproteinase-2. Biochemical and Biophysical Research Communications. 2000;**278**:390-394

[97] Fan SH, Wang YY, Lu J, Zheng YL, Wu DM, Zhang ZF, et al. CERS2 suppresses tumor cell invasion and is associated with decreased V-ATPase and MMP-2/MMP-9 activities in breast cancer. Journal of Cellular Biochemistry. 2015;**116**:502-513

[98] Wojtkowiak JW, Verduzco D, Schramm KJ, Gillies RJ. Drug resistance and cellular adaptation to tumor acidic pH microenvironment. Molecular Pharmaceutics. 2011;**8**:2032-2038

[99] Lu Q, Lu S, Huang L, Wang T, Wan Y, Zhou CX, et al. The expression of V-ATPase is associated with drug resistance and pathology of non-small-cell lung cancer. Diagnostic Pathology. 2013;**8**:145

[100] Kulshrestha A, Katara GK, Ibrahim S, Pamarthy S, Jaiswal MK, Gilman Sachs A, et al. Vacuolar ATPase 'a2' isoform exhibits distinct cell surface accumulation and modulates matrix metalloproteinase activity in ovarian cancer. Oncotarget. 2015;**6**:3797-3810

[101] Wiedmann RM, von Schwarzenberg K, Palamidessi A, Schreiner L, Kubisch R, Liebl J, et al. The V-ATPase-inhibitor archazolid abrogates tumor metastasis via inhibition of endocytic activation of the Rho-GTPase Rac1. Cancer Research. 2012;**72**:5976-5987

[102] Altan N, Chen Y, Schindler M, Simon SM. Defective acidification in human breast tumor cells and implications for chemotherapy. The Journal of Experimental Medicine. 1998;**187**:1583-1598

[103] Kim KH, Lee MS. Autophagy—A key player in cellular and body

metabolism. Nature Reviews. Endocrinology. 2014;**10**:322-337

[104] Gewirtz DA. The four faces of autophagy: Implications for cancer therapy. Cancer Research. 2014;**74**:647-651

[105] Kissing S, Hermsen C, Repnik U, Nesset CK, von Bargen K, Griffiths G, et al. Vacuolar ATPase in phagosome-lysosome fusion. The Journal of Biological Chemistry. 2015;**290**:14166-14180

[106] Mijaljica D, Prescott M, Devenish RJ. V-ATPase engagement in autophagic processes. Autophagy. 2011;7:666-668

[107] Carr G, Williams DE, Diaz-Marrero AR, Patrick BO, Bottriell H, Balgi AD, et al. Bafilomycins produced in culture by Streptomyces spp. isolated from marine habitats are potent inhibitors of autophagy. Journal of Natural Products. 2010;**73**:422-427

[108] Sorkin A, von Zastrow M. Endocytosis and signalling: Intertwining molecular networks. Nature Reviews. Molecular Cell Biology. 2009;**10**:609-622

[109] Marshansky V, Futai M. The V-type H+-ATPase in vesicular trafficking: Targeting, regulation and function. Current Opinion in Cell Biology. 2008;**20**:415-426

[110] Sun-Wada GH, Wada Y. Role of vacuolar-type proton ATPase in signal transduction. Biochimica et Biophysica Acta. 2015;**1847**:1166-1172

[111] Windler SL, Bilder D. Endocytic internalization routes required for delta/notch signaling. Current Biology. 2010;**20**:538-543

[112] Le Borgne R. Regulation of notch signalling by endocytosis and endosomal sorting. Current Opinion in Cell Biology. 2006;**18**:213-222

[113] Baron M. Endocytic routes to notch activation. Seminars in Cell & Developmental Biology. 2012;**23**:437-442

[114] Sorensen EB, Conner SD. Gamma-secretase-dependent cleavage initiates notch signaling from the plasma membrane. Traffic. 2010;**11**:1234-1245

[115] Lange C, Prenninger S, Knuckles P, Taylor V, Levin M, Calegari F. The H(+) vacuolar ATPase maintains neural stem cells in the developing mouse cortex. Stem Cells and Development. 2011;**20**:843-850

[116] Wada Y, Sun-Wada GH. Positive and negative regulation of developmental signaling by the endocytic pathway. Current Opinion in Genetics & Development. 2013;**23**:391-398

[117] Barth JM, Kohler K. How to take autophagy and endocytosis up a notch. BioMed Research International. 2014;**2014**:960803

[118] Lee JH, McBrayer MK, Wolfe DM, Haslett LJ, Kumar A, Sato Y, et al. Presenilin 1 maintains lysosomal Ca(2+) homeostasis via TRPML1 by regulating vATPase-mediated lysosome acidification. Cell Reports. 2015;**12**:1430-1444

[119] Polakis P. Wnt signaling in cancer. Cold Spring Harbor Perspectives in Biology. 2012;**4**(5):a008052

[120] Sebio A, Kahn M, Lenz HJ. The potential of targeting Wnt/beta-catenin in colon cancer. Expert Opinion on Therapeutic Targets. 2014;**18**:611-615

[121] Baarsma HA, Konigshoff M, Gosens R. The WNT signaling pathway from ligand secretion to gene transcription: Molecular mechanisms and

pharmacological targets. Pharmacology & Therapeutics. 2013;**138**:66-83

[122] Ichihara A. (pro)renin receptor and vacuolar H(+)-ATPase. The Keio Journal of Medicine. 2012;**61**:73-78

[123] Gao C, Cao W, Bao L, Zuo W, Xie G, Cai T, et al. Autophagy negatively regulates Wnt signalling by promoting Dishevelled degradation. Nature Cell Biology. 2010;**12**:781-790

[124] Guillard M, Dimopoulou A, Fischer B, Morava E, Lefeber DJ, Kornak U, et al. Vacuolar H+-ATPase meets glycosylation in patients with cutis laxa. Biochimica et Biophysica Acta. 2009;**1792**:903-914

[125] Cao X, Yang Q, Qin J, Zhao S, Li X, Fan J, et al. V-ATPase promotes transforming growth factor-beta-induced epithelial-mesenchymal transition of rat proximal tubular epithelial cells. American Journal of Physiology. Renal Physiology. 2012;**302**:F1121-F1132

[126] Moschetta M, Reale A, Marasco C, Vacca A, Carratu MR. Therapeutic targeting of the mTOR-signalling pathway in cancer: Benefits and limitations. British Journal of Pharmacology. 2014;**171**:3801-3813

[127] Zoncu R, Bar-Peled L, Efeyan A, Wang S, Sancak Y, Sabatini DM. mTORC1 senses lysosomal amino acids through an inside-out mechanism that requires the vacuolar H(+)-ATPase. Science. 2011;**334**:678-683

[128] Kim YC, Guan KL. mTOR: A pharmacologic target for autophagy regulation. The Journal of Clinical Investigation. 2015;**125**:25-32

[129] Bar-Peled L, Sabatini DM. Regulation of mTORC1 by amino acids. Trends in Cell Biology. 2014;**24**:400-406

[130] Webb BA, Chimenti M, Jacobson MP, Barber DL. Dysregulated pH: A perfect storm for cancer progression. Nature Reviews. Cancer. 2011;**11**:671-677

[131] Collins MP, Forgac M. Regulation of V-ATPase assembly in nutrient sensing and function of V-ATPases in breast cancer metastasis. Frontiers in Physiology. 2018;**9**:902

[132] Cotter K, Liberman R, Sun-Wada G, Wada Y, Sgroi D, Naber S, et al. The a3 isoform of subunit a of the vacuolar ATPase localizes to the plasma membrane of invasive breast tumor cells and is overexpressed in human breast cancer. Oncotarget. 2016;**7**:46142-46157

[133] Avnet S, Di Pompo G, Lemma S, Salerno M, Perut F, Bonuccelli G, et al. V-ATPase is a candidate therapeutic target for Ewing sarcoma. Biochimica et Biophysica Acta. 2013;**1832**:1105-1116

[134] Xu J, Xie R, Liu X, Wen G, Jin H, Yu Z, et al. Expression and functional role of vacuolar H(+)-ATPase in human hepatocellular carcinoma. Carcinogenesis. 2012;**33**:2432-2440

[135] Chung C, Mader CC, Schmitz JC, Atladottir J, Fitchev P, Cornwell ML, et al. The vacuolar-ATPase modulates matrix metalloproteinase isoforms in human pancreatic cancer. Laboratory Investigation. 2011;**91**:732-743

[136] Smith GA, Howell GJ, Phillips C, Muench SP, Ponnambalam S, Harrison MA. Extracellular and luminal pH regulation by vacuolar H+-ATPase isoform expression and targeting to the plasma membrane and endosomes. The Journal of Biological Chemistry. 2016;**291**:8500-8515

[137] Nishisho T, Hata K, Nakanishi M, Morita Y, Sun-Wada GH, Wada Y, et al.

The a3 isoform vacuolar type H(+)-ATPase promotes distant metastasis in the mouse B16 melanoma cells. Molecular Cancer Research. 2011;**9**:845-855

[138] von Schwarzenberg K, Wiedmann RM, Oak P, Schulz S, Zischka H, Wanner G, et al. Mode of cell death induction by pharmacological vacuolar H+-ATPase (V-ATPase) inhibition. The Journal of Biological Chemistry. 2013;**288**:1385-1396

[139] Schempp CM, von Schwarzenberg K, Schreiner L, Kubisch R, Muller R, Wagner E, et al. V-ATPase inhibition regulates anoikis resistance and metastasis of cancer cells. Molecular Cancer Therapeutics. 2014;**13**:926-937

[140] Lu X, Chen L, Chen Y, Shao Q, Qin W. Bafilomycin A1 inhibits the growth and metastatic potential of the BEL-7402 liver cancer and HO-8910 ovarian cancer cell lines and induces alterations in their microRNA expression. Experimental and Therapeutic Medicine. 2015;**10**:1829-1834

[141] Zhang S, Schneider LS, Vick B, Grunert M, Jeremias I, Menche D, et al. Anti-leukemic effects of the V-ATPase inhibitor Archazolid a. Oncotarget. 2015;**6**:43508-43528

[142] Vukovic V, Tannock IF. Influence of low pH on cytotoxicity of paclitaxel, mitoxantrone and topotecan. British Journal of Cancer. 1997;**75**:1167-1172

[143] Mahoney BP, Raghunand N, Baggett B, Gillies RJ. Tumor acidity, ion trapping and chemotherapeutics. I. Acid pH affects the distribution of chemotherapeutic agents in vitro. Biochemical Pharmacology. 2003;**66**:1207-1218

[144] Kulshrestha A, Katara GK, Ibrahim SA, Riehl V, Sahoo M, Dolan J, et al. Targeting V-ATPase isoform restores cisplatin activity in resistant ovarian cancer: Inhibition of autophagy, endosome function, and ERK/MEK pathway. Journal of Oncology. 2019;**2019**:2343876

[145] Marquardt D, Center MS. Involvement of vacuolar H(+)-adenosine triphosphatase activity in multidrug resistance in HL60 cells. Journal of the National Cancer Institute. 1991;**83**:1098-1102

[146] Licon-Munoz Y, Fordyce CA, Hayek SR, Parra KJ. V-ATPase-dependent repression of androgen receptor in prostate cancer cells. Oncotarget. 2018;**9**:28921-28934

[147] Steeg PS. Targeting metastasis. Nature Reviews. Cancer. 2016;**16**:201-218

[148] Mohamed MM, Sloane BF. Cysteine cathepsins: Multifunctional enzymes in cancer. Nature Reviews. Cancer. 2006;**6**:764-775

[149] Vasiljeva O, Papazoglou A, Kruger A, Brodoefel H, Korovin M, Deussing J, et al. Tumor cell-derived and macrophage-derived cathepsin B promotes progression and lung metastasis of mammary cancer. Cancer Research. 2006;**66**:5242-5250

[150] Withana NP, Blum G, Sameni M, Slaney C, Anbalagan A, Olive MB, et al. Cathepsin B inhibition limits bone metastasis in breast cancer. Cancer Research. 2012;**72**:1199-1209

[151] Capecci J, Forgac M. The function of vacuolar ATPase (V-ATPase) a subunit isoforms in invasiveness of MCF10a and MCF10CA1a human breast cancer cells. The Journal of Biological Chemistry. 2013;**288**:32731-32741

[152] McGuire CM, Collins MP, Sun-Wada G, Wada Y, Forgac M. Isoform-specific gene disruptions reveal a role for the V-ATPase subunit a4 isoform in the invasiveness of 4T1-12B breast cancer cells. The Journal of Biological Chemistry. 2019;**294**:11248-11258

[153] Di Cristofori A, Ferrero S, Bertolini I, Gaudioso G, Russo MV, Berno V, et al. The vacuolar H+ ATPase is a novel therapeutic target for glioblastoma. Oncotarget. 2015;**6**:17514-17531

[154] Liu P, Chen H, Han L, Zou X, Shen W. Expression and role of V1A subunit of V-ATPases in gastric cancer cells. International Journal of Clinical Oncology. 2015;**20**:725-735

[155] Manolson MF, Yu H, Chen W, Yao Y, Li K, Lees RL, et al. The a3 isoform of the 100-kDa V-ATPase subunit is highly but differentially expressed in large (>or=10 nuclei) and small (<or= nuclei) osteoclasts. The Journal of Biological Chemistry. 2003;**278**:49271-49278

[156] Feng S, Deng L, Chen W, Shao J, Xu G, Li YP. Atp6v1c1 is an essential component of the osteoclast proton pump and in F-actin ring formation in osteoclasts. The Biochemical Journal. 2009;**417**:195-203

[157] Lee SH, Rho J, Jeong D, Sul JY, Kim T, Kim N, et al. v-ATPase V0 subunit d2-deficient mice exhibit impaired osteoclast fusion and increased bone formation. Nature Medicine. 2006;**12**:1403-1409

[158] Lee BS, Holliday LS, Ojikutu B, Krits I, Gluck SL. Osteoclasts express the B2 isoform of vacuolar H(+)-ATPase intracellularly and on their plasma membranes. The American Journal of Physiology. 1996;**270**:C382-C388

[159] Tan LJ, Wang ZE, Wu KH, Chen XD, Zhu H, Lu S, et al. Bivariate genome-wide association study implicates ATP6V1G1 as a novel pleiotropic locus underlying osteoporosis and age at menarche. The Journal of Clinical Endocrinology and Metabolism. 2015;**100**:E1457-E1466

[160] Duan X, Liu J, Zheng X, Wang Z, Zhang Y, Hao Y, et al. Deficiency of ATP6V1H causes bone loss by inhibiting bone resorption and bone formation through the TGF-beta1 pathway. Theranostics. 2016;**6**:2183-2195

[161] Zhang Y, Huang H, Zhao G, Yokoyama T, Vega H, Huang Y, et al. ATP6V1H deficiency impairs bone development through activation of MMP9 and MMP13. PLoS Genetics. 2017;**13**:e1006481

[162] Komatsu M, Waguri S, Chiba T, Murata S, Iwata J, Tanida I, et al. Loss of autophagy in the central nervous system causes neurodegeneration in mice. Nature. 2006;**441**:880-884

[163] Hara T, Nakamura K, Matsui M, Yamamoto A, Nakahara Y, Suzuki-Migishima R, et al. Suppression of basal autophagy in neural cells causes neurodegenerative disease in mice. Nature. 2006;**441**:885-889

[164] Masters CL, Bateman R, Blennow K, Rowe CC, Sperling RA, Cummings JL. Alzheimer's disease. Nature Reviews Disease Primers. 2015;**1**:15056

[165] Lee JH, Yu WH, Kumar A, Lee S, Mohan PS, Peterhoff CM, et al. Lysosomal proteolysis and autophagy require presenilin 1 and are disrupted by Alzheimer-related PS1 mutations. Cell. 2010;**141**:1146-1158

[166] Poewe W, Seppi K, Tanner CM, Halliday GM, Brundin P, Volkmann J,

et al. Parkinson disease. Nature Reviews Disease Primers. 2017;**3**:17013

[167] Korvatska O, Strand NS, Berndt JD, Strovas T, Chen DH, Leverenz JB, et al. Altered splicing of ATP6AP2 causes X-linked parkinsonism with spasticity (XPDS). Human Molecular Genetics. 2013;**22**:3259-3268

[168] Gharanei S, Zatyka M, Astuti D, Fenton J, Sik A, Nagy Z, et al. Vacuolar-type H+-ATPase V1A subunit is a molecular partner of Wolfram syndrome 1 (WFS1) protein, which regulates its expression and stability. Human Molecular Genetics. 2013;**22**:203-217

[169] Ramser J, Abidi FE, Burckle CA, Lenski C, Toriello H, Wen G, et al. A unique exonic splice enhancer mutation in a family with X-linked mental retardation and epilepsy points to a novel role of the renin receptor. Human Molecular Genetics. 2005;**14**:1019-1027

[170] Hedera P, Alvarado D, Beydoun A, Fink JK. Novel mental retardation-epilepsy syndrome linked to Xp21.1-p11.4. Annals of Neurology. 2002;**51**:45-50

[171] Colacurcio DJ, Nixon RA. Disorders of lysosomal acidification-the emerging role of v-ATPase in aging and neurodegenerative disease. Ageing Research Reviews. 2016;**32**:75-88

[172] Brown D, Sabolic I, Gluck S. Polarized targeting of V-ATPase in kidney epithelial cells. The Journal of Experimental Biology. 1992;**172**:231-243

[173] Nelson RD, Guo XL, Masood K, Brown D, Kalkbrenner M, Gluck S. Selectively amplified expression of an isoform of the vacuolar H(+)-ATPase 56-kilodalton subunit in renal intercalated cells. Proceedings of the National Academy of Sciences of the United States of America. 1992;**89**:3541-3545

[174] Smith AN, Finberg KE, Wagner CA, Lifton RP, Devonald MA, Su Y, et al. Molecular cloning and characterization of Atp6n1b: A novel fourth murine vacuolar H+-ATPase a-subunit gene. The Journal of Biological Chemistry. 2001;**276**:42382-42388

[175] Stover EH, Borthwick KJ, Bavalia C, Eady N, Fritz DM, Rungroj N, et al. Novel ATP6V1B1 and ATP6V0A4 mutations in autosomal recessive distal renal tubular acidosis with new evidence for hearing loss. Journal of Medical Genetics. 2002;**39**:796-803

[176] Esmail S, Kartner N, Yao Y, Kim JW, Reithmeier RAF, Manolson MF. Molecular mechanisms of cutis laxa- and distal renal tubular acidosis-causing mutations in V-ATPase a subunits, ATP6V0A2 and ATP6V0A4. The Journal of Biological Chemistry. 2018;**293**:2787-2800

[177] Finberg KE, Wagner CA, Bailey MA, Paunescu TG, Breton S, Brown D, et al. The B1-subunit of the H(+) ATPase is required for maximal urinary acidification. Proceedings of the National Academy of Sciences of the United States of America. 2005;**102**:13616-13621

[178] Hennings JC, Picard N, Huebner AK, Stauber T, Maier H, Brown D, et al. A mouse model for distal renal tubular acidosis reveals a previously unrecognized role of the V-ATPase a4 subunit in the proximal tubule. EMBO Molecular Medicine. 2012;**4**:1057-1071

[179] Lorente-Canovas B, Ingham N, Norgett EE, Golder ZJ, Karet Frankl FE, Steel KP. Mice deficient in H+-ATPase a4 subunit have severe hearing impairment associated with enlarged endolymphatic compartments within the inner ear.

Disease Models & Mechanisms. 2013;**6**:434-442

[180] Berk DR, Bentley DD, Bayliss SJ, Lind A, Urban Z. Cutis laxa: A review. Journal of the American Academy of Dermatology. 2012;**66**(842):e1-e17

[181] Van Damme T, Gardeitchik T, Mohamed M, Guerrero-Castillo S, Freisinger P, Guillemyn B, et al. Mutations in ATP6V1E1 or ATP6V1A cause autosomal-recessive cutis Laxa. American Journal of Human Genetics. 2017;**100**:216-227

[182] Dou D, Revol R, Ostbye H, Wang H, Daniels R. Influenza a virus cell entry, replication, virion assembly and movement. Frontiers in Immunology. 2018;**9**:1581

[183] Hunt SR, Hernandez R, Brown DT. Role of the vacuolar-ATPase in Sindbis virus infection. Journal of Virology. 2011;**85**:1257-1266

[184] Chu JJ, Ng ML. Infectious entry of West Nile virus occurs through a clathrin-mediated endocytic pathway. Journal of Virology. 2004;**78**:10543-10555

[185] Minematsu A, Miyazaki T, Shimamura S, Nishikawa H, Nakayama H, Takazono T, et al. Vacuolar proton-translocating ATPase is required for antifungal resistance and virulence of Candida glabrata. PLoS One. 2019;**14**:e0210883

[186] Rane HS, Bernardo SM, Raines SM, Binder JL, Parra KJ, Lee SA. Candida albicans VMA3 is necessary for V-ATPase assembly and function and contributes to secretion and filamentation. Eukaryotic Cell. 2013;**12**:1369-1382

[187] Erickson T, Liu L, Gueyikian A, Zhu X, Gibbons J, Williamson PR. Multiple virulence factors of Cryptococcus neoformans are dependent on VPH1. Molecular Microbiology. 2001;**42**:1121-1131

Section 3

Interdisciplinary Approaches in Channel Targeting

Chapter 7

Structural Determinants for Ligand Accommodation in Voltage Sensors

Abigail García-Morales, Aylin López-Palestino and Daniel Balleza

Abstract

After ligand binding, many ion channels undergo rearrangements at the voltage sensor domain (VSD) that often modulate their gating activity with important physiological repercussions. Since the VSD is dynamic, it is interesting to establish a correlation between the potential mobility of this element in terms of its intrinsic flexibility and its ability to accommodate several ligands by induced-fit mechanisms. We presume that these associations are not causal since the flexibility of the VSD could have an important impact on the ligand coupling event. Many significantly flexible ion channels show a general architecture and composition compatible with important conformational changes and capable of accommodating chemically diverse agonists. In this contribution, the structural bases of this subtle and probably unexpected relationship between the VSD flexibility and its influence during the dynamic coupling of the ligand are exposed. Thus, given its physiological relevance, the study of ion channel malfunction can be associated with ligand accommodation events to the VSD, which could depend on its local flexibility. This could contribute to a better understanding of the molecular bases of a variety of physiological disorders. In consequence, considering these effects during the protein/ligand interaction could be determinant to the rational design of novel drugs.

Keywords: voltage sensor, side-chain flexibility, side-chain rotamers, RMSD, induced-fit docking, conformational selection

1. Introduction

Every cell is defined by its membrane. This amphiphilic molecular matrix is highly organized and complex in terms of its lipid and protein components. Ion channels have evolved to mediate ion transport throughout membranes always in favor of electrochemical gradients from free-living bacteria to mammal neurons [1, 2]. In doing so, these membrane proteins are part of complex physiological networks which include signal transduction processes, cellular communication, or the propagation of electrical signals [3]. The voltage-gated ion channel (VGIC) superfamily comprises dozens of variations on a common theme—(i) a voltage sensor domain (VSD) and (ii) a pore domain (PD) [4]. This modular architecture has in turn evolved into activation mechanisms as diverse as the detection of changes in the potential across

IntechOpen

the membrane, the binding of diverse chemical ligands, local membrane stretching, or subtle changes in temperature or the pH. In consequence, those physical-chemical variables are often interrelated modulating the ion channel gating but not clearly defined as exclusive stimuli for a determined protein [5]. In voltage sensing, the VSD performs important conformational rearrangements moving through the membrane-electric field and coupling this motion to the opening of the permeation pathway at the PD. To do this, four transmembrane segments (S1–S4) at the VSD respond sensing the electric field by translocating the so-called gating charges and by reorganization of the dipole moments into an aqueous crevice around the S4 segments, so that at any membrane potential, the charged side-chains of basic residues (mainly Arg and Lys) are essentially both hydrated and ionized either above or below the plane of the lipid bilayer (i.e. depolarized or hyperpolarized, respectively) [6].

In that sense, the VSD is clearly a mobile and intrinsically flexible element. A more detailed analysis of this mobility has revealed the relative flexibility of the different regions into this domain and clearly demonstrate that helices S1, S2, and the N-terminal part of S3 (S3a) are relatively more static than the so-called VSD *paddle* (S3b-S4) which loop linker show enhanced flexibility at higher temperatures in molecular simulations [7]. In those *in silico* studies, it is also evident that the segment S4 undergoes hinge bending and swiveling about its central axis, motion facilitated by the conformational instability of the S3a helix [7, 8]. The intrinsic

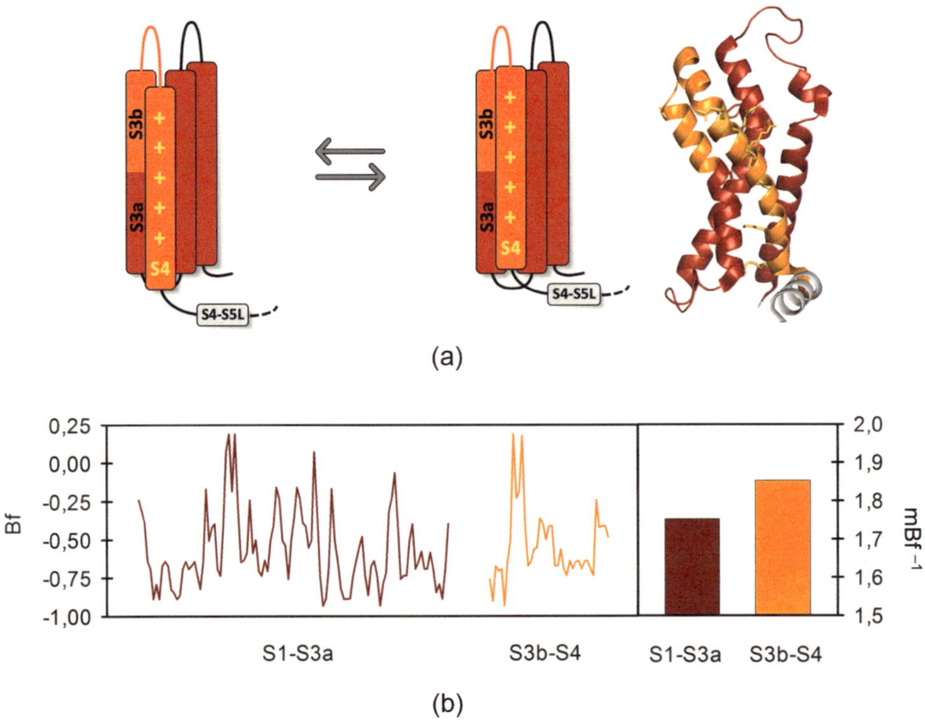

Figure 1.
Flexibility plot for the VSD of Kv1.2-Kv2.1 paddle chimera (PDB code: 2r9r) calculated using the method previously reported by us [5]. a. Cartoons showing the hyperpolarized (left) and depolarized (right) conformations of the VSD. b. Plots showing B-factors normalized to a Gumbel distribution according to our previous studies. Segment S3b-S4, involved in channel activation and the electromechanical coupling exhibits higher flexibility than the rest of the voltage sensor (S1–S3a).

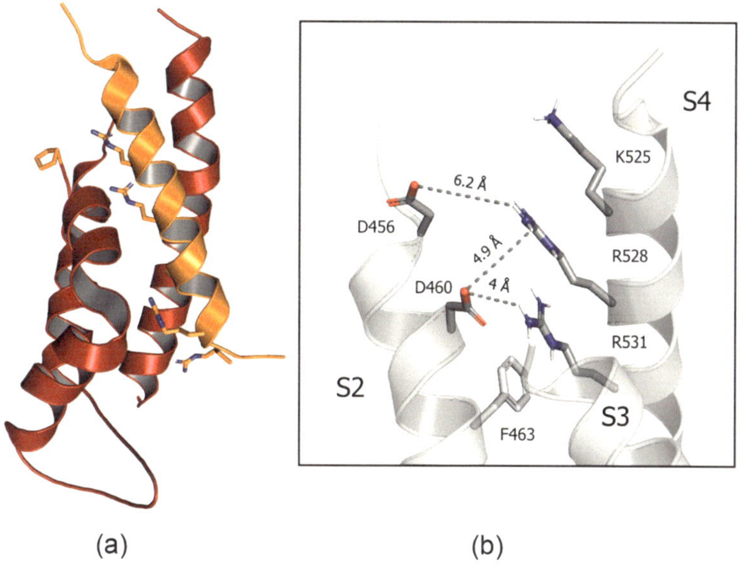

Figure 2.
The gating charge transfer center (CTC) in the voltage sensors of Kv channels. a. Ribbon representation of the four segments (S1–S4) making one VSD of the human ether-a-go-go-related gene (hERG) channel (PDB code: 7cno). b. Close-up view of the CTC highlighting the gating charges on the S4 segment. Side-chains of the positively charged residues on S4 (labeled as R and K) and negative residues D456 and D460 interact forming salt bridges. The hydrophobic residue F463 controls the energy barrier of the final gating transition. Figures containing structures were prepared with Pymol (http://www.pymol.org/).

flexibility of the S3a region facilitates movement of the segment S3b-S4, which in turn exhibits an even higher flexibility profile due to its composition rich in residues with small side-chain (Gly, Ser, Thr) and basic residues (**Figure 1**). Notably, these predictions have been experimentally confirmed elsewhere [9, 10] and some reports also indicate that abnormal S4 movements cause pathological effects related for example to the development of epilepsy [11]. Therefore, it is becoming increasingly clear that the VSD is a flexible dynamic structure with evident relevance in physiological disorders.

The mechanisms of gating in ion channels have been intensively studied. On activation, outward S4 motion is associated with specific interactions with conserved negative countercharges (Asp and Glu) in transmembrane segments S1, S2, and S3 by forming sequential salt bridges with the positively charged residues in S4 inside an aqueous pore (**Figure 2**). Such interactions facilitate the translocation of the S4 segment in an energetically unfavorable membrane environment promoting the sequential salt-bridge formation and the electromechanical activation of the S4-S5 linker, which directly couples voltage sensor movement to the activation gate [12]. These negative countercharges are well-conserved in S1, S2, and S3 transmembrane segments in Kv channels (**Figure 3**). Besides, several VSD countercharge mutations associated with disease phenotypes including neural, cardiac, or skeletal muscle disorders have also been identified [14]. Despite channelopathies often affect ion channel gating, many of these pathologies have yet to be functionally or biophysically characterized. In this regard and given the diverse physiological and pathophysiological functions played by members of the VGIC superfamily, the VSD becomes a promising target for rational drug design.

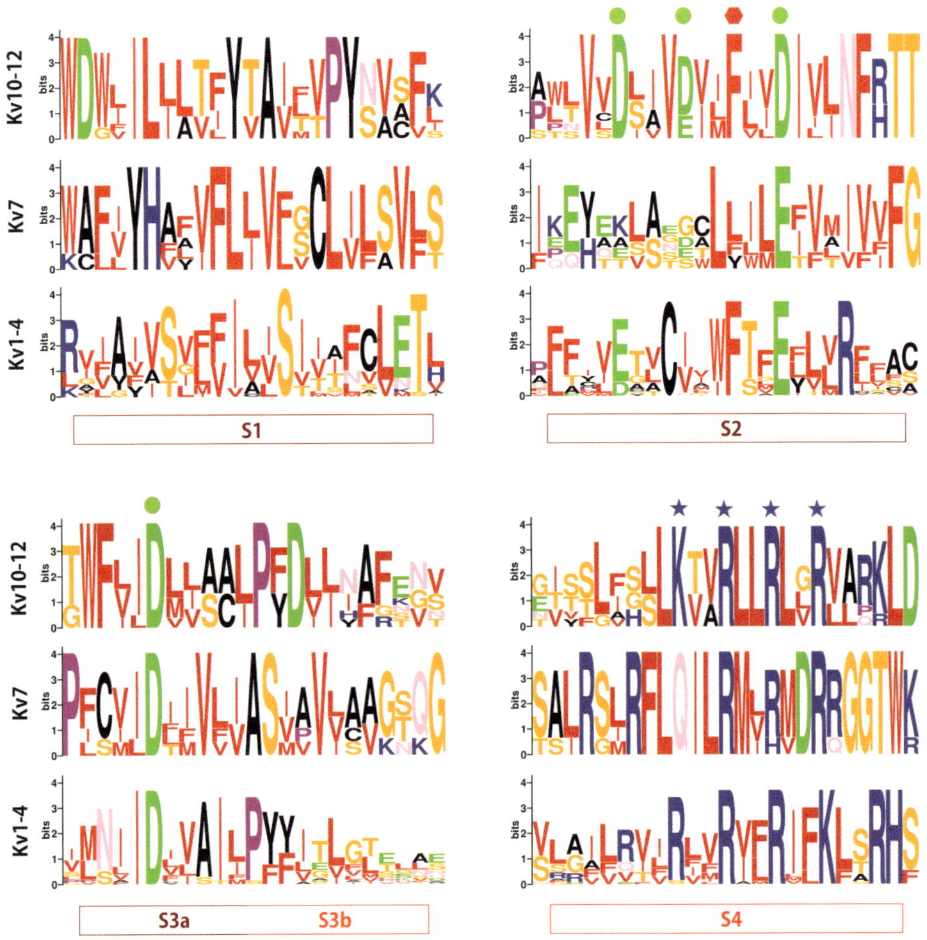

Figure 3.
Sequence logos from the three main subfamilies of voltage-gated potassium channels using the webserver WebLogo (http://weblogo.threeplusone.com/). The bars below the sequence logos represent the extent of transmembrane segments S1–S4. The consensus sequences generated are represented statistically, showing the relative conservation of each residue at that position. The height of each letter in every position indicates the maximum theoretical entropy for protein sequences (measured in bits) [13] which is determined by the number of aligned sequences and the degree of ambiguity in the alignments for each residue. Blue stars indicate basic residues involved in voltage sensing on S4. Green circles indicate conserved negative countercharges in S2 and S3. Red hexagon depicts a well-conserved aromatic residue that controls the transfer of the more inner gating charge, shaping the electric field inside the voltage sensor. Conserved residues addressed in this study are color-coded as: Red (hydrophobic); green (acidic); blue (basic); orange (small side-chain); pink (flexible side-chain); and black (rigid side-chain).

2. The voltage sensor domain as a pharmacological target

There are many reports on the interactions of different intracellular ligands with ion channels and particularly important is the well-understood interaction of ligands with the cytosolic tail domain (CTD) in large-conductance calcium-activated potassium channels (BK_{Ca}), which contain several binding sites. Also relevant are the studies of the interaction of cGMP or cAMP with the cyclic nucleotide-binding domain (CNBD) in cyclic nucleotide-gated (CNG) and hyperpolarization-activated (HCN) channels. Structural and functional information has shown that frequently the ligand-binding sites in those channels are clustered located at the interface between the cytosolic domain and the VSD, acting synergistically to activate the gate at the PD [15, 16]. In other studies, ligands have been found directly coupled to the VSD influencing the channel activation, opening, closing, or inactivating the pore, such as some protein toxins from tarantula do [17, 18] or the binding of vanilloids, monoterpenoids, and related compounds to the S1–S4 domain in the transient receptor potential (TRP) channels [19].

From a structural perspective, the study of interactions between specific chemical ligands and the VSD in ion channels opens the possibility to rationally design both agonist and antagonist drugs. Let us have a look at three specific cases—(1) the voltage- and lipid-gated potassium channel KCNQ2 (**Figure 4**), (2) the

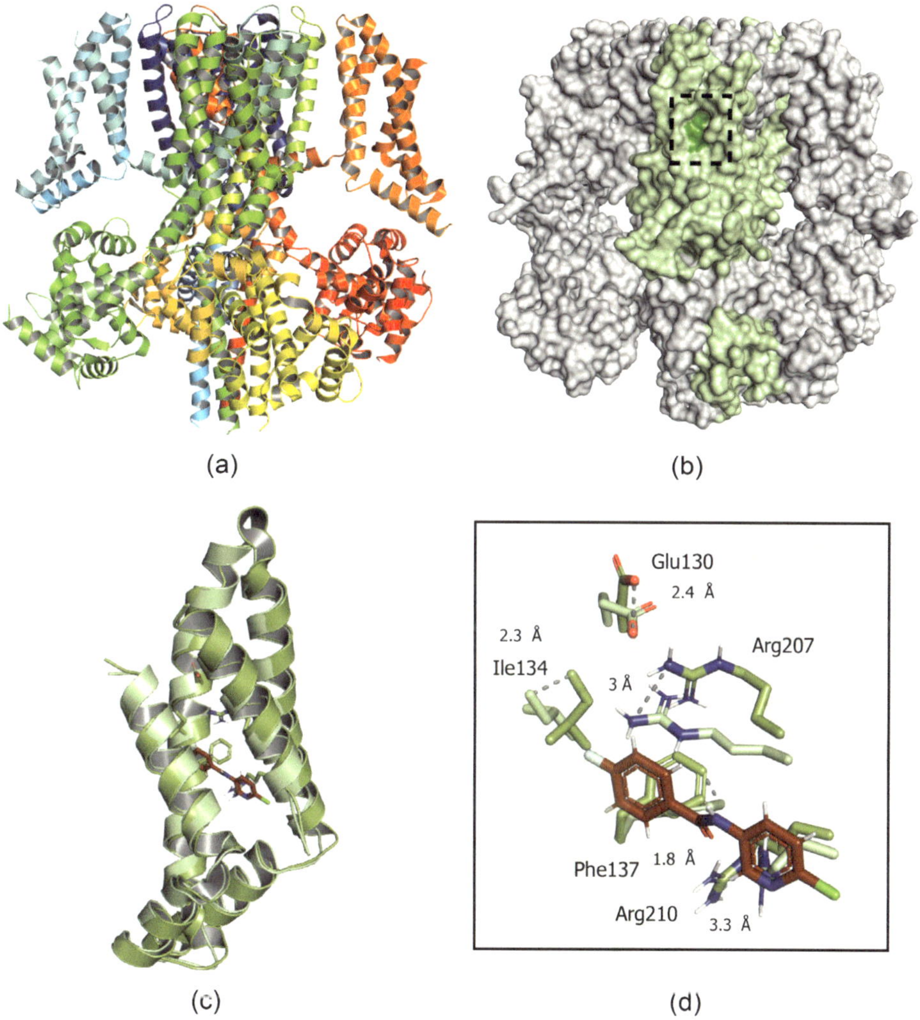

Figure 4.
Structure of the human Kv7.2 (KCNQ2) channel (PDB accessions codes 7cr0 (apostate) and 7cr1 (in complex with ztz240). a. Ribbon representation of side view showing the VSDs exposed to the lipid bilayer (not shown here). b. Surface representation showing the ligand-binding pocket as a box. c. the unliganded structure of the VSD (highlighted in pale green) is superimposed with the one in complex with ligand (highlighted in smudge green). d. Three-dimensional stick representation of the Kv7.2 pocket in complex with ligand ztz240 (dark brown). The distances traveled by the side-chain conformational rotamers are shown as dashes. The overlapping residues correspond to the side-chains involved in ligand binding.

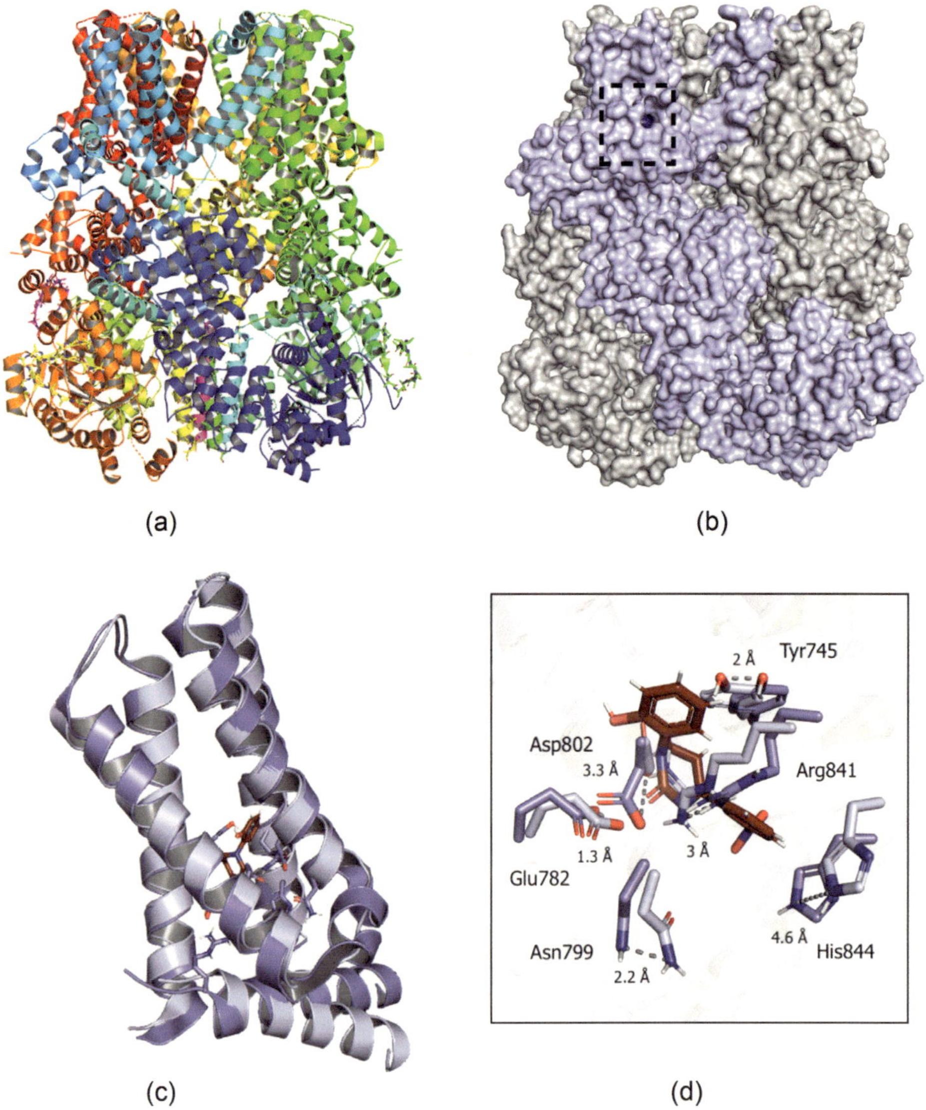

Figure 5.
*Structure of the TRPM8 ion channel from the collared flycatcher (Ficedula albicollis) (PDB codes 6bpq (apostate) and 6nr3 (in complex with icilin, PI(4,5)P2, and Ca^{2+}). a. Ribbon representation of side view. b. Surface representation showing the ligand-binding pocket as a box. c. The unliganded structure of the VSLD is highlighted in light blue and superimposed with the one in complex with icilin (semi-dark blue). d. Overlay of both structures to visualize the side-chain conformation changes (see **Figure 3** for details).*

cold/menthol activated TRPM8 channel (**Figure 5**), and (3) the capsaicin receptor TRPV1 (**Figure 6**). Studies on the KCNQ2 (Kv7.2) potassium channel show that the aromatic amide ztz240, a derivative of niclosamide, binds to the open configuration of the VSD. This interaction directly couples such a chemical ligand with a binding pocket of 170 $Å^3$ located between some specific residues at segments S2 (Glu130, Ile134, Phe137) and S4 (Arg207 and Arg210), i.e. precisely in the gating charge pathway of this ion channel [20]. This raises the possibility to design drugs using the channel gating pore in voltage-dependent channels as a therapeutic target [21]. In this case, ztz240 and some derived chemotypes have demonstrated important

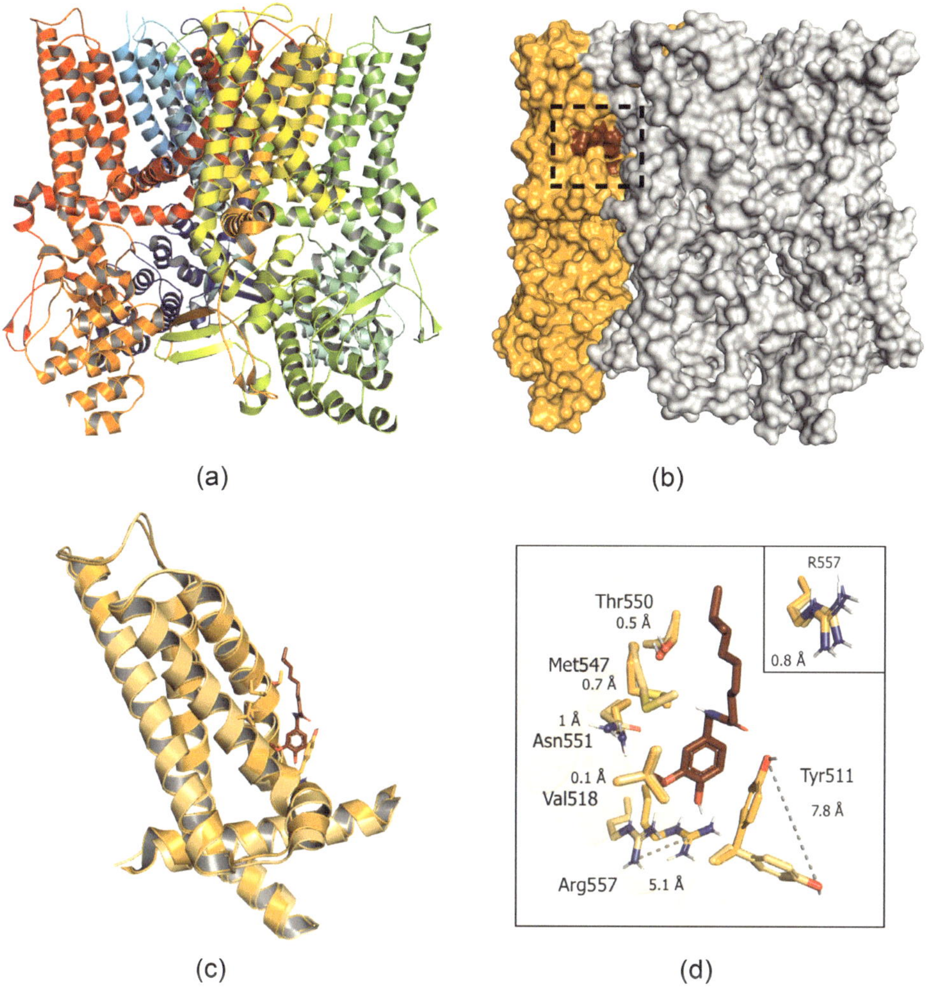

Figure 6.
Structure of the TRPV1 channel from rat (PDB codes 7lp9 (apostate, 4°C), 7lpe (in complex with capsaicin, 48°C), and 7lpa (in complex with capsaicin, 4°C). a. Side view of the tetramer in ribbon representation. b. Surface representation showing the ligand-binding pocket as a box. c. The unliganded structure of the VSLD is highlighted in light orange and superimposed with the one in complex with capsaicin (bright orange). d. Overlay of structures at 4°C (apostate) and 48°C (in presence of capsaicin) to visualize two different rotamers by residue (see ***Figure 3*** *for details). Inset: Superposition of Arg557 side-chain obtained at 4°C for apo- and holo-structures.*

anti-epileptic activity, which might be valuable for the treatment of epilepsy (see below). Remarkably, this interaction in the gating charge pathway of KCNQ2, considering the induced-fit model, demonstrates that this pocket may accommodate different activators [20].

In the case of TRPM8, an analog binding pocket has been described as highly adaptable to accommodate diverse chemical structures in distinct orientations. Both agonists and antagonists are dynamically recognized in this promiscuous pocket making the whole S1–S4 domain conformationally dynamic and transmitting these rearrangements to the TRP helix, but without inducing important changes in its overall structure [22]. Menthol, the main compound of mint, is the clue activator to understand how TRPM8 channels are ligand-activated. It binds to the cavity formed between S1 and S4 by a so-called "grab and stand" mechanism. The hydroxyl group of menthol works as a hand to specifically grab with Arg842 (segment S4) through a hydrogen bond, while its isopropyl "legs" stand on residues on S4 through electrostatic interactions. Thus, menthol binding induced widespread conformational rearrangements in the S1–S4 domain which open the S6 bundle gate to allow ion permeation [23].

On the other hand, since TRPV1 channels participate in several pathways of neuronal inflammatory signaling, it also represents an attractive therapeutic target for the treatment of neuroinflammation, neurodegenerative diseases, and chronic pain. Feng et al. [24] have studied diverse diarylurea compounds by molecular docking and dynamics, finding that specific residues located in the interface between the VSD and the PD are implicated in several van der Waals interactions. Particularly important are residues Tyr511, Leu518, Leu547, Thr550, Asn551, Arg557, and Leu670. Besides these observations, residues at the base of the interface between the VSD and the PD (segments S3, S4, S5, and the S4-S5 linker) are important binding sites for N-(3-fluoro-4-methylsulfonamidomethylphenyl)urea. Docking analysis of this compound with human TRPV1 has revealed that hydrogen bonding and π–π interactions with Tyr511 (segment S3) and hydrophobic interactions with two pockets in the S3 and S4 segments (residues Met514, Leu515 and Leu547, Thr550, respectively) are critical for its activity. Flexible docking studies have also revealed that N-benzyl phenylsulfonamide derivatives of 2-(3-fluoro-4- methylsulfonamidophenyl)propanamide specifically bind to the same pockets, being again critical for the potent activity of these antagonists [25, 26]. Notably, conformational analyses have revealed that the S1–S4 domain in TRPV channels remains relatively static during opening [27–29].

These three examples are mechanistically different, but they all share something in common, and this is the specific association of their respective ligands at the VSD, as well as their direct association with the potential difference across the membrane. However, even though TRP channels have frequently been treated as strictly ligand-dependent, it is increasingly clear that their voltage sensitivity could be underestimated [30]. These reports support the idea that the increasingly available detailed structural information, as well as detailed functional studies, greatly simplifies the search for chemical modulators with agonistic or antagonistic action. In consequence, the identification of potential ligand-binding sites in the VSD makes the rational design of new drugs, the goal of several research efforts.

2.1 Side-chain flexibility and ligand accommodation

Proteins are intrinsically flexible. This property derives from the two bonds associated with the carbon α (Cα), which can freely rotate and contribute to the flexibility of the main backbone. The torsion angles Phi (Φ) and Psi (ψ) represent the rotation around the Cα-N bond and the one around the Cα-carbonyl bond, respectively. However, this rotational capacity also depends on the steric and conformational effects of the associated side-chains. In these terms, one classical structural parameter to estimate the mobility of each atom into a protein structure is the so-called B-factor, which reflects the degree of thermal motion and static disorder. This parameter, also called the Debye–Waller factor, represents the atomic displacement of the macromolecule and is used in protein crystallography to describe the attenuation of X-ray or neutron scattering caused by thermal motion, which reflects the uncertainty in atom positions [31, 32]. Therefore, proteins are intrinsically rigid or flexible ultimately based on their primary sequence.

From this perspective, in addition to their known physical and chemical properties, if their relative location parameters are considered, amino acids can also be classified in terms of their contribution to the flexibility of a given segment within a protein. The amino acids that are considered rigid generally consist of those that exhibit bulky side-chains, generally cyclic or aromatic, those that have heavy heteroatoms, and are generally hydrophobic. On the other hand, amino acids considered flexible are those having polar side-chains, have charges, or whose side group is only a proton, i.e. glycine [33]. Proline deserves a separate discussion, as this amino acid has been considered both rigid and flexible in terms of its kinking effect on alpha helices [34].

Typically, flexibility in proteins has been visualized in terms of local and global motions, which include—(i) the multiple conformations that a certain residue can acquire in the polypeptide chain, (ii) local-scale fluctuations in the conformation of the side chains with respect to the backbone, and (iii) massive movements of subdomains with respect to another part of the protein [35]. In this last regard, a pioneer study of protein crystal structures shows that intrinsic flexibility can be distinguished in terms of hinge motions and shear displacements between close-packed segments of the protein [36]. It is becoming clearer that studying protein flexibility and the multiple side-chain conformations during molecular docking is very relevant since it may contribute to a favorable change in the Gibbs binding free energy by optimizing the van der Waals interactions between the protein and the ligand. This favors a change in enthalpy and minimizes the decrease in entropy [37, 38], albeit protein flexibility also depends on several other factors, including heat capacity, conformational entropy, salt bridge networks, electrostatic interactions, and the hydrophobic effect [39]. In any case, the study of side-chain flexibility in ion channels and how it contributes to ligand accommodation could be critical to understand molecular recognition events and predict ligand binding. This could open novel therapeutic strategies for the treatment of diverse neuropathic disorders.

2.2 Three cases of study: Kv7.2, TRPM8, TRPV1

Since flexible regions in proteins can be predicted from the primary sequence through the evaluation of the normalized B-factor for a determined structure [40], we have implemented an easy algorithm in Excel based on the procedure carried out by Smith and cols. [32, 33]. In general terms, B-factor normalization, Bn, depends on (1) the atomic thermal factor, B, reported on the PDB, (2) the sample mean value of B-factors, Bm, for a dataset of protein structures, and (3) the standard deviation of the sample distribution of such factors, $B\sigma$, for a determined structure [41]. In sum, normalized B-factors are indicative of each local residue flexibility that can be calculated as:

$$Bn = \frac{B - Bm}{B\sigma} \tag{1}$$

Based on these theoretical principles, we implemented an algorithm to predict local side-chain flexibility, which correlates the composition of amino acids in a protein sequence in the context of its N- and C-terminal neighbors. The program assigns a weighted normalized B-factor value based on a stiffness classification for each amino acid, in accordance with previously published results [33]. This software, so-called FlexiProt, includes Trp, Tyr, Phe, Cys, Ile, Val, His, Leu, and Met as rigid amino acids and the rest of them, *i.e.* Gly, Thr, Arg, Ser, Asn, Gln, Asp, Pro, Glu, and Lys as flexible. The program incorporates a graphical generator of local flexibility profiles, based on primary sequence, to help the user better visualize this disorder parameter for its subsequent structural evaluation. Flexiprot is friendly and interactive since although it runs automatically, it allows at all times the possibility of evaluating local subsequences for a better prediction of the internal flexibility parameter, based on structural aspects associated with the theoretical degrees of freedom of the side-chains.

Through this type of sequence analysis, we show in **Figure 7** the local flexibility profiles of three distinct channels for the segments S3 to S4—(1) human KCNQ2 (hKv7.2), (2) mouse TRPM8 (mTRPM8), and (3) rat TRPV1 (rTRPV1). As in the Kv1.2-2.1 *paddle* chimera (**Figure 1**), in these three cases, local flexibility for segment S4 is always higher than the one for segment S3. However, the profile of the S3 segment in the TRPV1 channel is significantly stiffer than the other two, which have similar flexibility profiles for the same segment, although the S4 segment of the voltage-dependent channel Kv7.2 is considerably more flexible compared to their thermosensitive counterparts. This notorious flexibility is mainly determined by the highly conserved Arg residues, responsible for transporting the gating charges in the voltage sensor during the activation mechanism of these proteins [5]. Considering the new structures available for this channel in the presence of the ztz240 modulator, whose binding site is precisely in the intimacy of the voltage sensor, it becomes interesting to analyze the role of side-chain flexibility for each of the interacting

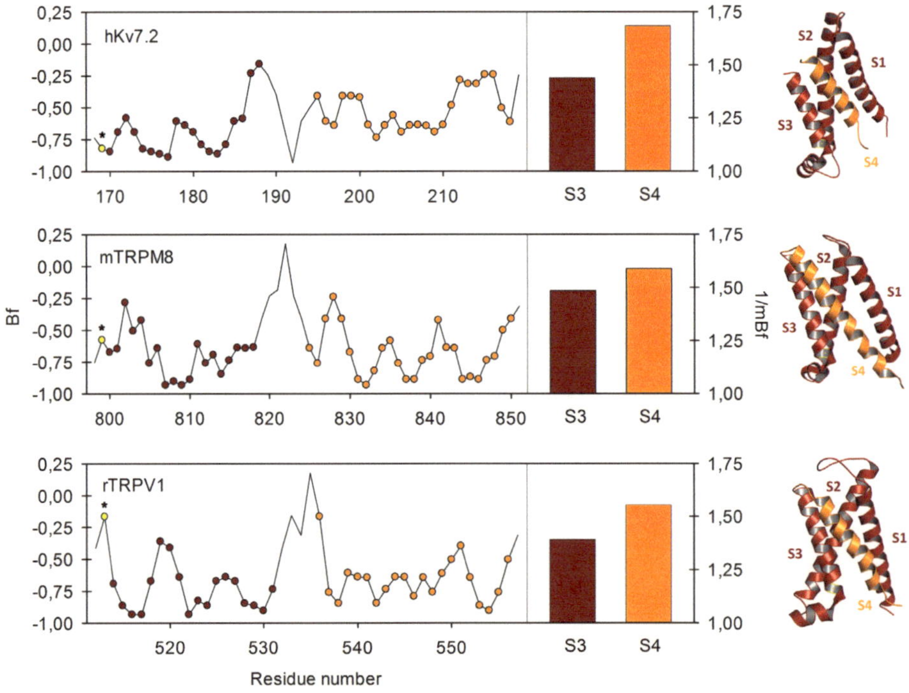

Figure 7.
Predicted flexibility for segments S3 and S4 in human Kv7.2, TRPM8 (mouse), and TRPV1 (rat). Flexibility (defined by B-factor values) using the FlexiProt algorithm according to Ref. [5] shows a higher flexibility profile for segment S4 compared to S3. Predictions also indicate that the three channels follow the flexibility ranking TRPV1 < TRPM8 < Kv7.2. Asterisk indicates the start of the first turn of the α-helix at the N-terminal part of segment S3a. mBf: mean B-factor.

residues and that has been mentioned elsewhere [20, 42]. **Figure 4** show this interaction in two conformational states for the VSD of the Kv7.2 channel, in the absence and presence of the state-dependent modulator ztz240. For a segment of 145 residues that encompass the integrity of the VSD, a root-mean-square deviation (RMSD) of 1.159 Å is indicative of a fine accommodation for this drug without representing a significant conformational change of this protein domain. It is clearly noted that the intrinsically flexible residues Arg207 and Arg 210 at the S4 segment undergo an important conformational rearrangement that allows the ligand to be adequately accommodated through van der Waals interactions. The analysis of the structures also indicates that these residues are displaced 3.0 and 3.3 Å respectively. Another important residue, implicated in the potentiation of the activity of ztz240 is Glu130, an amino acid considered even more flexible than Arg [33], which is found as a countercharge in the S2 segment and whose side group moves 2.4 Å during the interaction. In contrast to these data, two other amino acids also important for the interaction, Ile134 and Phe137—amino acids of a rigid nature—show a rearrangement of 2.3 Å and 1.8 Å, respectively. These data suggest that the significant local flexibility of the S4 segment in this channel strongly contributes to ligand accommodation with minor effects on the large movements that the VSD experience during activation and that some of the main residues interact with this ligand move in a range of 1.8–3.3 Å.

In TRPM8, a channel described also as sensitive to voltage [43, 44], a similar effect to Kv7.2/ztz240 is observed. According to recent structural studies of this channel, icilin, a compound derived from tetrahydropyrimidine-2-one and more potent than menthol as the agonist, binds to the voltage-sensor-like domain (VSLD) mainly through van der Waals interactions to residues Tyr745 (S1), Glu782 (S2) Asn799 (S3), Asp802 (S3), Arg841 (S4), and His844 (S4) [28, 45, 46]. Analogously to that seen in the Kv7.2 channel, the significant flexibility exhibited by the S3 and S4 segments in the TRPM8 channel contributes to a fine accommodation of icilin through conformational rearrangements of these amino acids in a range from 2 to 4.6 Å (**Figure 5D**). These displacements occur in the context of an RMSD of 0.89 Å over 123 Cα atoms which integrate the VSLD of this cold-sensitive channel. Similarly, the interaction of the antagonist TC-I 2014 in the same cavity of the VSLD [28] induces small rearrangements of the corresponding side-chains implicated in ligand sensitivity, with displacements of around 1–3 Å and an average RMSD of 0.427 Å (data not shown). Taken together, these observations suggest that the high flexibility profile in the S3 and primarily the S4 segments of these transmembrane domains facilitates a fine repositioning of the participating side-chains, which are implicated in ligand accommodation.

The case of the transient receptor potential vanilloid subtype 1 channel is slightly different. **Figure 6** shows the interaction of capsaicin with the VSLD of TRPV1. Thanks to the recently released structural data of TRPV1 channels in presence of this ligand, a more detailed exploration of such interactions as a function of the associated local flexibility contribution of the VSLD contributes to a better understanding of this process. For a segment of 166 residues encompassing the whole VSLD and part of the TRPbox, an RMSD of 0.63 Å suggest an almost null conformational change for this part of the protein in the course of the closed-to-open transition during the ligand interaction, as it has been previously reported [25]. In this case, residues Val518, Met547, Thr550, and Asn551 move their side-chains less than 1 Å when they interact with capsaicin, whereas Tyr511 and Arg557 experience a significant displacement of 7.8 Å and 5.1 Å, respectively. These observations are consistent with the low predicted local flexibility for segment S3 in this channel (**Figure 7**, **Table 1**) [5]. Furthermore, they are also in good agreement with the vision that the VSLD acts as a rigid body during TRPV1 activation [25].

Channel	S1–S4 length (Cα)	S3 (mBf)	S4 (mBf)	RMSD (Å)	Side-chain displacement (Å)
Kv7.2	145	1.44	1.68	1.16	2.4 (E130) 2.3 (I134) 1.8 (F137) 3.0 (R207) 3.3 (R210)
TRPM8	123	1.48	1.59	0.89	2.0 (Y745) 1.3 (E782) 2.2 (N799) 3.3 (D802) 4.6 (H844)
TRPV1	166	1.39	1.55	0.63	7.8 (Y511) 0.1 (V518) 0.7 (M547) 0.5 (T550) 5.1 (R557)

mBf, mean B-factor (1/mBf).

Table 1.
Flexibility parameters for the studied channels upon ligand interaction.

Our analysis shows that in this case, the low flexibility profile of the S3 segment contributes to creating a rigid crevice. This structure accommodates the catechol/vanilloid ring of capsaicin at the base of the VSLD where the bulky side-chain of Tyr511 residue rearranges with a displacement of almost 8 Å during a transition from 4 to 48°C. It is very significant that this tyrosine, which frequently is classified as a rigid side-chain with low conformational entropy, in the context of the structure of this channel, carries out a significant rearrangement even greater than Arg557, which has been frequently quantified as much more flexible [33, 47]. Besides, the side-chain of Arg557 at S4 undergoes a conformational rearrangement of 5.1 Å during the interaction but an almost null displacement (0.8 Å) if this is carried out at 4°C (**Figure 6D**, *inset*). Indeed, despite its low flexibility, tyrosine has been considered very effective for mediating molecular recognition maybe because changing the orientation of its side-chain from *gauche* negative (g–) to *trans* (t) conformation is equivalent to moving the hydroxyl group around 9 Å, which is the length of an average drug molecule [48, 49]. On the other hand, the motion of Arg557 is associated with the formation of a hydrogen bond with the Glu570 residue on the S4-S5 linker, leading to its swivel [50]. This also confirms that arginine often participates in molecular recognition events. The terminal positively charged guanidinium group of this residue affords multiple geometries due to its long side-chain can retain substantial residual conformational entropy occupying several rotameric states, while maintaining specific interactions through its charged functional group [51]. In sum, we hypothesize that, in contrast with the mechanism dependent on the large local (S3–S4) flexibility of the voltage-dependent Kv7.2 channel or the cold/menthol-activated TRPM8 channel, these large conformational changes in specific residues compensate for the low mobility that the whole transmembrane domain (i.e. the VSLD), as a rigid body, undergoes during TRPV1 activation.

3. Side-chain flexibility and the dynamic nature of protein-ligand interactions

Despite the large increase in deposition of crystallographic, NMR, and cryo-electron microscopy structures in recent years, little dynamic information regarding the conformational degrees of freedom of protein structures is currently available. *In silico* local flexibility theoretical prediction together with molecular dynamics algorithms are likely to be useful in helping to solve this limitation. Diverse computational strategies have been developed to explore the side-chain rotameric states as a function of the primary sequence, backbone structure, and ligand interaction by molecular docking in specific protein motifs [52–54]. Besides, the prediction of protein flexibility [32, 33, 40, 41], as well as its identification and visualization [55], have been a constant goal in protein research. Side-chain flexibility represents an intrinsic property of amino acids, as it correlates with configurational entropy differences and indeed is related to the generation of dynamic rotamers, which are defined as a particular combination of side-chain dihedral angles [38].

Given the dynamic and multifactorial nature of flexibility in proteins, trying to predict the binding mode of any ligand is an inspiring challenge. However, the use of predictive tools, dynamic simulations, and specific experimental tests will facilitate a better understanding of the molecular mechanisms underlying ligand-dependent modulation of ion channels. This could be of great impact on the rational design and discovery of novel drugs. Therefore, in the case of the study of the VSD as a ligand-binding motif, side-chain flexibility is especially relevant and must be always considered in light of the induced-fit model and conformational selection mechanisms [56]. In these terms, since side-chain and also frequently the backbone are subject to rearrangements upon ligand interaction (*see* our previous analysis above), we suggest that any experimental approach to develop novel drugs should be designed from the perspective of a dynamic target. In this sense, the study by Li and cols. is very relevant since they identify a hydrophobic pocket inside the charge transfer center (CTC) of the Kv7.2 channel which can accommodate different chemical ligands [20]. This enables such openers to regulate ion channel activation and offers new therapeutic strategies for the treatment of several hyperexcitability disorders, such as epilepsy and neuropathic pain. Since the VSD exhibits important conformational freedom during the gating process, it is important to note that this class of ligands preferentially binds to specific conformational states. The compound ztz240, for example, is accommodated to a hydrophobic pocket only when the VSD is in its activated conformation. This interaction stabilizes the activated state of the channel, thus contributing to its antiepileptic activity [57]. Therefore, it is conceivable to consider the so-called gating pore as an important target for ion channel research given its potential adaptability to new openers or inhibitors.

In the **Figure 8** included at the end of this study, the workflow for the development of drugs with therapeutic potential is shown sequentially. *In silico* and structural studies, from the perspective of the "induced fit" model and the "conformational selection" hypothesis [58, 59], both contribute to a better understanding of the dynamic aspects of the protein/ligand interactions. In these terms, the role of side-chain flexibility becomes pivotal, and methods to predict it, such as the one performed herein, as well as methods to visualize it, such as those reported elsewhere [55] are

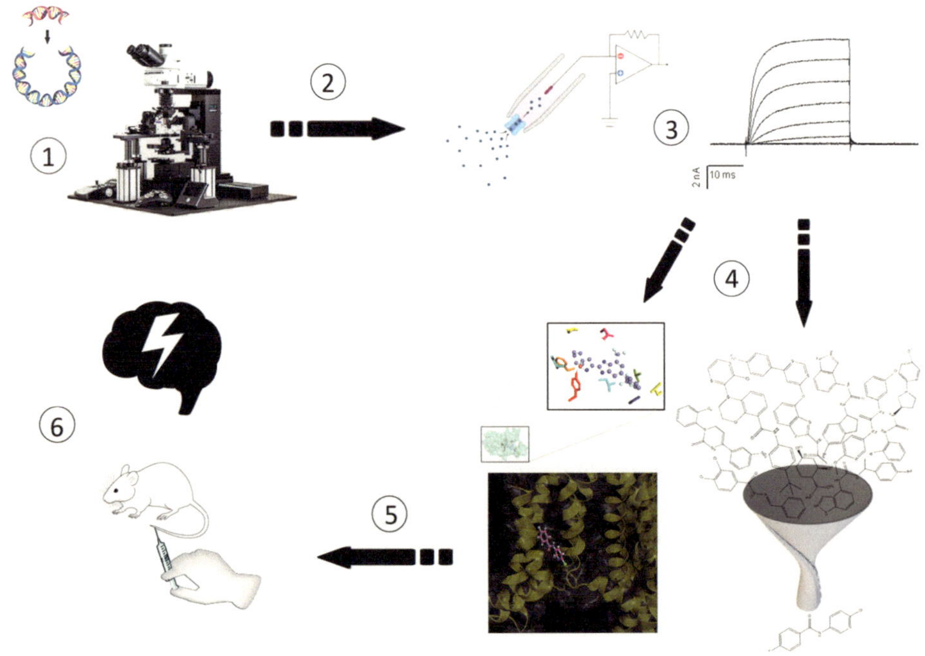

Figure 8.
Compounds that act on the voltage sensor could be a good alternative for the development of new analgesic drugs and provide a complement to pain therapy. After synthesis of a compound with pharmacological potential, ion channel mutagenesis (1) and associated electrophysiological tests (2) are performed. The study of the activation/ inactivation properties of ionic channels with therapeutic interest (3) is decisive for establishing correlations between the adaptability of the molecular target and the candidate drug (4). Considering side-chain intrinsic flexibility and the degree of pocket disorder during these molecular recognition events allows the identification of residues crucial for drug activity. Finally, the new active compounds are tested for their in vivo validation using animal models (5). This strategy could be applied to the discovery of several modulators capable of dealing with diverse neurological disorders (6).

indispensable analysis tools. By the appropriate selection of chemical candidates with pharmacological potential, ion channels with defined mutations can be experimentally evaluated, shedding some light on the involvement of specific residues in ligand accommodation and their effects on voltage sensor regulation. Furthermore, thanks to the correct study of the flexibility profiles for a given segment, it is also possible to evaluate the nature of these interactions, providing additional information about the degrees of freedom necessary for an adequate ligand accommodation. With this background, experimental testing in animal models and eventual clinical studies becomes an achievable goal for the treatment of neurological disorders and problems of acute pain.

From this perspective, the predictive analysis of local flexibility that we have performed here on three different ion channels clearly shows how the characteristic ligands of each protein are accommodated in the binding sites generating important conformational changes in the side-chains of specific residues. In a very revealing way, we found that the Kv7.2 channel and the TRPM8 have a VSD and a VSLD with high S3-S4 flexibility profiles (mBf of 1.44/1.48 for S3 and 1.68/1.59 for S4, respectively) (**Table 1**). These domains show small local conformational changes according to the corresponding RMSD values calculated (1.16 Å and 0.89 Å respectively), while on the other hand, a channel such as TRPV1, whose flexibility profile is rather rigid (mBf = 1.39 for S3 and 1.55 for S4) exhibits a still minor conformational change

(RMSD = 0.63 Å for the equivalent segment) during the interaction with its specific ligand. Likewise, the conformational changes that we observe in the participating side-chains are also revealing, since the rotamers generated during the ligand interaction in flexible VSDs (Kv7.2 and TRPM8) are the result of rotations less than 3 Å on average, while in the case of the rigid S1-S4 segment of the TRPV1 channel, the conformational changes of the side-chains are less than 1 Å but two residues, in particular, Tyr511 (S3) and Arg557 (S4), undergo rotations of 7.8 Å and 5.1 Å, respectively, which suggest a different mechanism for ligand accommodation. This is even more revealing when it is considered that the conformational changes observed in the side-chain of Arg557 were obtained at 48°C [48] while the conformational rotamer for that residue at 4°C is less than 1 Å and that corresponding to Tyr 511 is even higher (8.2 Å) with a side-chain angle rotation of ~101°, which also suggest that motion of this residue is critical for ligand binding (**Figure 6D**).

After this analysis, we speculate that two different mechanisms for ligand accommodation in ion channels exist, which seem to be dependent on the conformational freedom of the VSD. These mechanisms could be interpreted as VSDs that have higher degrees of freedom, i.e. flexible and more prone to fine induced-fit mechanisms, which adapt better to the ligand through small displacements of multiple participating side-chains. On the other hand, more rigid VSDs follow the classical lock and key model for enzyme-substrate interactions, in which the ligand is accommodated directly but with important conformational changes in certain very specific residues. The conformational freedom of these specific residues would compensate for the low mobility observed in the rest of the structure.

4. Conclusions

Flexible regions in proteins are critical elements for the recognition of macromolecular interactions and induced molecular flexibility is essential to understand the principles of molecular recognition between ligand and receptor. However, the nature of side-chain flexibility is elusive and dynamic processes involving this flexible component are among the most difficult to characterize. Given its direct participation in the activation of voltage-dependent channels, the voltage-sensing domain is a very attractive target from the therapeutic point of view. As side-chain flexibility represents an intrinsic property of amino acids which is correlated with configurational entropy differences, it is now known that rotamer changes in specific residues during ligand interaction are finely synchronized [38]. Our analysis has confirmed this claim. According to our predictive algorithm, the local flexibility in S3–S4 segments which are implied to ligand binding in three different channels, correlated well with the adaptability of specific residues through the generation of side-chain rotamers during each interaction. However, what is intriguing is the fact that segments exhibiting high flexibility profiles (i.e. Kv7.2 and TRPM8) are correlated with small-scale changes and the generation of side-chain rotamers that are more homogeneously and subtly accommodated among the participating residues during ligand binding, while those which are slightly more rigid (i.e. TRPV1) remain practically immobile during the interaction, except for one or two residues that undergo a very pronounced conformational rearrangement to accommodate the drug. Since it is assumed that these new conformations are energetically favorable states [60], in terms of drug design these observations might not be trivial when considering the induced-fit model [58, 59] in combination with the conformational selection hypothesis [61].

In agreement with them, the dynamic binding of drugs to a specific protein target may lead to chemoselectivity, high ligand affinity as well as the favoring of long residence times in the binding site. There are many potential interactions if those factors are considered and reasonably well understood. In these terms, the identification of side-chain rotamer rearrangements upon ligand binding in combination with the use of the global RMSD comparison between two protein conformers is more informative. Thus, the incorporation of predictive tools of side-chain flexibility in protein/ligand interactions is key to infer dynamic aspects in molecular docking. Besides, the experimental evaluation of new drugs from this perspective becomes pivotal in the rational design of therapeutic strategies to control several physiological disorders and face emerging channelopathies.

Acknowledgements

The authors wish to thank the Division of Postgraduate Studies and Research (DEPI) of the Technological Center of Mexico, Veracruz campus, for the facilities for conducting this study.

Notes

FlexiProt 2.0 is a software designed for the prediction of flexibility profiles in primary sequences. Today, our group is working to share it in the public domain. For more information or in case of interest, contact the corresponding author at: daniel.bm@veracruz.tecnm.mx.

Author details

Abigail García-Morales, Aylin López-Palestino and Daniel Balleza*
Tecnológico Nacional De México/IT-Veracruz, Veracruz, Mexico

*Address all correspondence to: daniel.bm@veracruz.tecnm.mx

References

[1] Balleza D, Quinto C, Elias D, Gómez-Lagunas F. A high-conductance cation channel from the inner membrane of the free-living soil bacteria *Rhizobium etli*. Archives of Microbiology. 2010;**192**(7):595-602. DOI: 10.1007/s00203-010-0587-3

[2] Labarca P, Bacigalupo J. Ion channels from chemosensory olfactory neurons. Journal of Bioenergetics and Biomembranes. 1988;**20**(5):551-569. DOI: 10.1007/BF00768919

[3] Hille B, Armstrong C, MacKinnon R. Ion channels: From idea to reality. Nature Medicine. 1999;**5**:1105-1109. DOI: 10.1038/13415

[4] Yu FH, Yarov-Yarovoy V, Gutman GA, Catterall WA. Overview of molecular relationships in the voltage-gated ion channel superfamily. Pharmacological Reviews. 2005;**57**:387-395. DOI: 10.1124/pr.57.4.13

[5] Balleza D, Rosas ME, Romero-Romero S. Voltage vs. Ligand I: Structural basis of the intrinsic flexibility of S3 segment and its significance in ion channel activation. Channels (Austin). 2019;**13**(1):455-476. DOI: 10.1080/19336950.2019.1674242

[6] Horn R. How ion channels sense membrane potential. Proceedings of the National Academy of Sciences of the United States of America. 2005;**102**(14):4929-4930. DOI: 10.1073/pnas.0501640102

[7] Sands ZA, Grottesi A, Sansom MS. The intrinsic flexibility of the Kv voltage sensor and its implications for channel gating. Biophysical Journal. 2006;**90**(5):1598-1606. DOI: 10.1529/biophysj.105.072199

[8] Jensen MØ, Jogini V, Borhani DW, Leffler AE, Dror RO, Shaw DE. Mechanism of voltage gating in potassium channels. Science. 2012;**336**(6078):229-233. DOI: 10.1126/science.1216533

[9] Kalstrup T, Blunck R. S4–S5 linker movement during activation and inactivation in voltage-gated K+ channels. Proceedings of the National Academy of Sciences of the United States of America. 2018;**115**:E6751-E6759

[10] Lee CH, MacKinnon R. Voltage sensor movements during hyperpolarization in the HCN channel. Cell. 2019;**179**(7):1582-1589.e7. DOI: 10.1016/j.cell.2019.11.006

[11] Pantazis A, Kaneko M, Angelini M, Steccanella F, Westerlund AM, Lindström SH, et al. Tracking the motion of the Kv1.2 voltage sensor reveals the molecular perturbations caused by a *de novo* mutation in a case of epilepsy. The Journal of Physiology. 2020;**598**(22):5245-5269. DOI: 10.1113/JP280438

[12] Long SB, Campbell EB, Mackinnon R. Voltage sensor of Kv1.2: Structural basis of electromechanical coupling. Science. 2005;**309**(5736):903-908. DOI: 10.1126/science.1116270

[13] Schneider TD, Stephens RM. Sequence logos: A new way to display consensus sequences. Nucleic Acids Research. 1990;**18**:6097-6100. DOI: 10.1093/nar/18.20.6097

[14] Groome JR, Bayless-Edwards L. Roles for countercharge in the voltage sensor domain of ion channels. Frontiers in Pharmacology. 2020;**11**:160. DOI: 10.3389/fphar.2020.00160

[15] Yang H, Zhang G, Cui J. BK channels: Multiple sensors, one activation gate.

Frontiers in Physiology. 2015;**6**:29. DOI: 10.3389/fphys.2015.00029

[16] James ZM, Zagotta WN. Structural insights into the mechanisms of CNBD channel function. The Journal of General Physiology. 2018;**150**(2):225-244. DOI: 10.1085/jgp.201711898

[17] Alabi AA, Bahamonde MI, Jung HJ, Kim JI, Swartz KJ. Portability of *paddle* motif function and pharmacology in voltage sensors. Nature. 2007;**450**(7168):370-375. DOI: 10.1038/nature06266

[18] Bosmans F, Martin-Eauclaire MF, Swartz KJ. Deconstructing voltage sensor function and pharmacology in sodium channels. Nature. 2008;**456**(7219):202-208. DOI: 10.1038/nature07473

[19] Steinberg X, Lespay-Rebolledo C, Brauchi S. A structural view of ligand-dependent activation in thermoTRP channels. Frontiers in Physiology. 2014;**5**:171. DOI: 10.3389/fphys.2014.00171

[20] Li P, Chen Z, Xu H, Sun H, Li H, Liu H, et al. The gating charge pathway of an epilepsy-associated potassium channel accommodates chemical ligands. Cell Research. 2013;**23**(9):1106-1118. DOI: 10.1038/cr.2013.82

[21] Kornilov P, Peretz A, Attali B. Channel gating pore: A new therapeutic target. Cell Research. 2013;**23**(9):1067-1068. DOI: 10.1038/cr.2013.89

[22] Diver MM, Cheng Y, Julius D. Structural insights into TRPM8 inhibition and desensitization. Science. 2019;**365**(6460):1434-1440. DOI: 10.1126/science.aax6672

[23] Xu L, Han Y, Chen X, Aierken A, Wen H, Zheng W, et al. Molecular mechanisms underlying menthol binding and activation of TRPM8 ion channel. Nature Communications. 2020;**11**(1):3790. DOI: 10.1038/s41467-020-17582-x

[24] Feng Z, Pearce LV, Zhang Y, Xing C, Herold BK, Ma S, et al. Multi-functional diarylurea small molecule inhibitors of TRPV1 with therapeutic potential for neuroinflammation. The AAPS Journal. 2016;**18**(4):898-913. DOI: 10.1208/s12248-016-9888-z

[25] Ann J, Sun W, Zhou X, Jung A, Baek J, Lee S, et al. Discovery of N-(3-fluoro-4-methylsulfonamidomethylphenyl)urea as a potent TRPV1 antagonistic template. Bioorganic & Medicinal Chemistry Letters. 2016;**26**(15):3603-3607. DOI: 10.1016/j.bmcl.2016.06.010

[26] Ann J, Ki Y, Yoon S, Kim MS, Lee JU, Kim C, et al. 2-Sulfonamidopyridine C-region analogs of 2-(3-fluoro-4-methylsulfonamidophenyl)propanamides as potent TRPV1 antagonists. Bioorganic & Medicinal Chemistry. 2016;**24**(6):1231-1240. DOI: 10.1016/j.bmc.2016.01.051

[27] Cao E, Liao M, Cheng Y, Julius D. TRPV1 structures in distinct conformations reveal activation mechanisms. Nature. 2013;**504**(7478):113-118. DOI: 10.1038/nature12823

[28] Singh AK, McGoldrick LL, Sobolevsky AI. Structure and gating mechanism of the transient receptor potential channel TRPV3. Nature Structural & Molecular Biology. 2018;**25**(9):805-813. DOI: 10.1038/s41594-018-0108-7

[29] McGoldrick LL, Singh AK, Saotome K, Yelshanskaya MV, Twomey EC, Grassucci RA, et al. Opening of the human epithelial calcium channel TRPV6. Nature. 2018;**553**(7687):233-237. DOI: 10.1038/nature25182

[30] Nilius B, Talavera K, Owsianik G, Prenen J, Droogmans G, Voets T. Gating of TRP channels: A voltage connection? The Journal of Physiology. 2005; **567**(Pt 1):35-44. DOI: 10.1113/jphysiol. 2005.088377

[31] Sun Z, Liu Q, Qu G, Feng Y, Reetz MT. Utility of B-factors in protein science: Interpreting rigidity, flexibility, and internal motion and engineering thermostability. Chemical Reviews. 2019;**119**(3):1626-1665. DOI: 10.1021/acs. chemrev.8b00290

[32] Radivojac P, Obradovic Z, Smith DK, Zhu G, Vucetic S, Brown CJ, et al. Protein flexibility and intrinsic disorder. Protein Science. 2004;**13**(1):71-80. DOI: 10.1110/ ps.03128904

[33] Smith DK, Radivojac P, Obradovic Z, Dunker AK, Zhu G. Improved amino acid flexibility parameters. Protein Science. 2003;**12**(5):1060-1072. DOI: 10.1110/ ps.0236203

[34] Williams KA, Deber CM. Proline residues in transmembrane helices: Structural or dynamic role? Biochemistry. 1991;**30**(37):8919-8923. DOI: 10.1021/bi00101a001

[35] Sánchez-Martinez M. Protein flexibility: From local to global motions. A computational study [PhD thesis]. Universitat de Barcelona; 2014. DOI: 10.13140/2.1.2411.0883

[36] Gerstein M, Lesk AM, Chothia C. Structural mechanisms for domain movements in proteins. Biochemistry. 1994;**33**(22):6739-6749. DOI: 10.1021/ bi00188a001

[37] Zavodszky MI, Kuhn LA. Side-chain flexibility in protein-ligand binding: The minimal rotation hypothesis. Protein Science. 2005;**14**(4):1104-1114. DOI: 10.1110/ps.041153605

[38] Gaudreault F, Chartier M, Najmanovich R. Side-chain rotamer changes upon ligand binding: Common, crucial, correlate with entropy and rearrange hydrogen bonding. Bioinformatics. 2012;**28**(18):i423-i430. DOI: 10.1093/bioinformatics/bts395

[39] Karshikoff A, Nilsson L, Ladenstein R. Rigidity versus flexibility: The dilemma of understanding protein thermal stability. The FEBS Journal. 2015;**282**:3899-3917. DOI: 10.1111/ febs.13343

[40] Schlessinger A, Rost B. Protein flexibility and rigidity predicted from sequence. Proteins. 2005;**61**(1):115-126. DOI: 10.1002/prot.20587

[41] Carugo O, Argos P. Correlation between side-chain mobility and conformation in protein structures. Protein Engineering. 1997;**10**(7):777-787. DOI: 10.1093/protein/10.7.777

[42] Wang CK. Complex mechanisms of KCNQ channel activation by state-dependent modulators [master of science thesis]. University of Alberta; 2018. p. 86

[43] Voets T, Owsianik G, Janssens A, Talavera K, Nilius B. TRPM8 voltage sensor mutants reveal a mechanism for integrating thermal and chemical stimuli. Nature Chemical Biology. 2007;**3**(3):174-182. DOI: 10.1038/ nchembio862

[44] Fernández JA, Skryma R, Bidaux G, Magleby KL, Scholfield CN, McGeown JG, et al. Voltage- and cold-dependent gating of single TRPM8 ion channels. The Journal of General Physiology. 2011;**137**(2):173-195. DOI: 10.1085/jgp.201010498

[45] Yin Y, Le SC, Hsu AL, Borgnia MJ, Yang H, Lee SY. Structural basis of cooling agent and lipid sensing by the cold-activated TRPM8 channel. Science.

2019;**363**(6430):eaav9334. DOI: 10.1126/science.aav9334

[46] Huang Y, Fliegert R, Guse AH, Lü W, Du J. A structural overview of the ion channels of the TRPM family. Cell Calcium. 2020;**85**:102111. DOI: 10.1016/j.ceca.2019.102111

[47] Najmanovich R, Kuttner J, Sobolev V, Edelman M. Side-chain flexibility in proteins upon ligand binding. Proteins. 2000;**39**(3):261-268. DOI: 10.1002/(sici)1097-0134(20000515)39:3<261::aid-prot90>3.0.co;2-4

[48] Koide S, Sidhu SS. The importance of being tyrosine: Lessons in molecular recognition from minimalist synthetic binding proteins. ACS Chemical Biology. 2009;**4**(5):325-334. DOI: 10.1021/cb800314v

[49] Källblad P, Dean PM. Efficient conformational sampling of local side-chain flexibility. Journal of Molecular Biology. 2003;**326**(5):1651-1665. DOI: 10.1016/s0022-2836(03)00083-4

[50] Kwon DH, Zhang F, Suo Y, Bouvette J, Borgnia MJ, Lee SY. Heat-dependent opening of TRPV1 in the presence of capsaicin. Nature Structural & Molecular Biology. 2021;**28**(7):554-563. DOI: 10.1038/s41594-021-00616-3

[51] Trbovic N, Cho JH, Abel R, Friesner RA, Rance M, Palmer AG 3rd. Protein side-chain dynamics and residual conformational entropy. Journal of the American Chemical Society. 2009;**131**(2):615-622. DOI: 10.1021/ja806475k

[52] Leach AR. Ligand docking to proteins with discrete side-chain flexibility. Journal of Molecular Biology. 1994;**235**:345-356. DOI: 10.1016/s0022-2836(05)80038-5

[53] Antunes DA, Devaurs D, Kavraki LE. Understanding the challenges of protein flexibility in drug design. Expert Opinion on Drug Discovery. 2015;**10**(12):1301-1312. DOI: 10.1517/17460441.2015.1094458

[54] Liu H, Lin F, Yang JL, Wang HR, Liu XL. Applying side-chain flexibility in motifs for protein docking. Genomics Insights. 2015;**8**:1-10. DOI: 10.4137/GEI.S29821

[55] Palamini M, Canciani A, Forneris F. Identifying and visualizing macromolecular flexibility in structural biology. Frontiers in Molecular Biosciences. 2016;**3**:47. DOI: 10.3389/fmolb.2016.00047

[56] Stank A, Kokh DB, Fuller JC, Wade RC. Protein binding pocket dynamics. Accounts of Chemical Research. 2016;**49**(5):809-815. DOI: 10.1021/acs.accounts.5b00516

[57] Li X, Zhang Q, Guo P, Fu J, Mei L, Lv D, et al. Molecular basis for ligand activation of the human KCNQ2 channel. Cell Research. 2021;**31**(1):52-61. DOI: 10.1038/s41422-020-00410-8

[58] Koshland DE. Application of a theory of enzyme specificity to protein synthesis. Proceedings of the National Academy of Sciences of the United States of America. 1958;**44**(2):98-104. DOI: 10.1073/pnas.44.2.98

[59] Boehr DD, Nussinov R, Wright PE. The role of dynamic conformational ensembles in biomolecular recognition. Nature Chemical Biology. 2009;**5**(11):789-796. DOI: 10.1038/nchembio.232

[60] Dunbranck RL, Cohen FE. Bayesian statistical analysis of protein side-chain rotamer preferences. Protein Science. 1997;**6**(8):1661-1681. DOI: 10.1002/pro.5560060807

[61] Rubin MM, Changeux JP. On the nature of allosteric transitions: Implications of non-exclusive ligand binding. Journal of Molecular Biology. 1966;**21**(2):265-274. DOI: 10.1016/0022-2836(66)90097-0